20.35

D1515274

SOLAR ENERGY IN AMERICA

The SCIENCE Report on

SOLAR ENERGY IN AMERICA

William D. Metz
Allen L. Hammond

 **American Association for the Advancement of Science**

Library of Congress Cataloging in Publication Data

Metz, William D.
    Solar energy in America

    Bibliography: p.
    Includes indexes and appendixes.
    1. Solar energy — Technology — Research I. Hammond,
Allen L., joint author. II. Science. III. Title
TJ810.M48   621.47   78-69957
ISBN 0-87168-301-6
ISBN 0-87168-238-9 pbk.

AAAS Publication 78-10

Part of the material in this book originally appeared as a series in the Research News
section of SCIENCE, the Journal of the American Association for the Advancement of
Science. The authors are members of the Research News staff of SCIENCE.

© 1978 American Association for the Advancement of Science
        1515 Massachusetts Avenue, NW
        Washington, D.C. 20005

Printed in the United States of America

Contents

Figures

Introduction

Solar energy is a subject that was hardly heard of before the prospect of scarce oil was thrust upon us, but it is undergoing a popular and technical revival that has few parallels in recent history.

The signs of renewed interest in solar energy are many. The federal research budget has burgeoned from $1 million to over $400 million per year since the 1973 embargo of oil to the United States. Knowledge that was buried in a few classic texts five years ago has become know-how pamphleteered by consumer groups, and information about solar products is being disbursed over telephone hot lines by federal outreach programs. Thousands of state and congressional legislators now take an active interest in solar issues, and industrial giants, as well as hundreds of smaller companies, are rushing to get solar products onto the commercial market. The supposed archenemies of solar power, the oil companies, are playing as large a role as any. No fewer than five major companies—Exxon, Mobil, ARCO, Shell and the major French company (Compagnie Française des Pétroles)—are investing heavily in advanced research on solar cells.

There are many reasons for the resurgence of interest in solar energy. It offers a new way of doing things at a time when many people are disenchanted with the old. It promises renewable energy when importing depletable energy is seen as a national liability. It promises clean energy when the degradation of the environment is becoming readily apparent. It traces its origins to peacetime research at a time when the major alternatives have developed as a by-product of nuclear weapons research. It offers individuals a degree of energy independence unprecedented in the industrial age, and it has sufficient potential for changing the institutional structures of the country so that some at the centers of power find it threatening. Solar energy is embraced by large numbers of consumer and reformist groups—more than other energy lobbies, its support has some of the attributes of a political movement.

The characteristics of solar energy responsible for its broad appeal are not accidents of time or place, but can be traced to the nature of the solar resource, which is unusually versatile. The technologies that derive their energy from the sun in one way or another are as distinctively different as windmills and wood stoves, solar cells and factory ships, solar furnaces and alcohol plantations. Systems that collect sunlight directly constitute only one set of solar technologies—the direct technologies. Wind, waves, plant growth, warming of tropical oceans and the terrestrial rain cycle are also driven by the flux of solar energy that falls on the earth, and they provide the basis for another set of solar technologies that have completely different characteristics from each other and from the direct technologies. If each technology—direct and indirect—is examined more closely, it can be seen to include a whole family of related technical concepts, each of which has different advantages and challenges. For wind energy, for instance, there are dozens of different approaches, and the other solar technologies are similarly fertile. Indeed, the breadth and richness of the technologies that derive their energy from the sun is one of the strongest arguments for optimism about the role that solar energy will eventually play in the national energy picture.

An indication of the speed of solar development is that almost all the solar technologies emphasized today are being developed around concepts that have emerged in the past five years. Only one of the technologies discussed in this book bears any conceptual resemblance to the best ideas of 1973, and that one has expanded and progressed faster than most authorities dreamed possible. Most solar technologies are still too expensive to be economic. But at a time when the development of nuclear energy has come to a virtual standstill, fusion is undergoing serious reexamination, and synthetic fuels from coal are making only halting progress, the vitality of solar research is one of the most hopeful signs for the future.

Exactly how much energy solar can contribute depends upon how you define it. Broadly defined, solar is already the source of a significant amount of energy for the United States. In 1977, plant matter (often called biomass) provided half as much energy as nuclear power did, largely through the burning of wood wastes at pulp and paper mills to produce energy to run those plants. Hydroelectric facilities—which derive their energy from the sun-driven rain cycle—provided slightly more energy than nuclear power did in 1977. Taking these two sources together, solar energy now provides about 5 percent of the U.S. energy needs.

How much energy solar can provide in the future depends on which solar technologies you bet on to prove economic and desirable. Some solar options are mature technologies today, others could make it to the marketplace very quickly with a little government subsidy or luck in the laboratory, while still others will require decades of development to reach the commercial stage, if they do so at all. There is a large disparity in solar energy supply projections. Forecasters who project minimal solar contributions generally assume that the long-term technologies will be the future solar workhorses, while those who project substantial contributions generally assume that the mature technologies will carry the day. Yet the number of mature solar technologies is greater than generally realized, and there is no technical reason why they cannot begin to be used at once. The solar projections for 2000 range from a low of 2 percent of U.S. needs (not counting hydropower as solar) to a high of 25 percent. The high figure may be optimistic, but it is easily possible that solar energy—broadly defined—could provide 10 to 15 percent of the country's energy by the end of the century and considerably more in following decades. More specific projections from a variety of sources are given in the first appendix.

One of the most distinctive aspects of solar energy is its potential for use in decentralized applications. Because of its flexibility, diversity, and variety of scale, solar energy can be put to use in novel ways. Solar energy supplies could therefore be structured very differently from the mainstays of today's energy economy.

The federal program for developing solar resources has been painfully slow to recognize these special characteristics, and has overwhelmingly emphasized solar systems that could be plugged into the energy network as exact replacements of older systems that are taken out. Since the material in this book initially appeared, there have been a number of changes in the government program. But the twin assumptions that traditional values should determine the shape of new energy systems and that "bigger is better" seem to persist.

In late 1977, the solar research program was reorganized as part of the creation of the Department of Energy, and the core of solar research was split into two or possibly three components, with most of the research and development (R&D) programs under one assistant secretary and the home and industrial process heating programs under another. Another reorganization, which is just being completed as this book goes to press, further divides the R&D programs into centralized and decentralized categories. The result is a program much more fragmented than it was two years ago. So far there is no indication—especially in the important matter of budget priorities—that solar energy will now be accorded equal status with nuclear energy or coal development, or that the basic philosophy behind the solar program has changed. The most innovative ideas continue to come from outside the government program and federal sponsorship continues to favor large-scale technologies. Ironically, federal energy officials seem to have answered outside criticism by decentralizing the solar program rather than the technology it is supposed to promote.

In spite of all the attractive attributes of solar energy, a warning is in order. Unforeseen problems could arise. Just as nuclear power appeared in the early stages of its development to be an ideal energy solution, and was in fact advocated by the Sierra Club for a period in the 1960s, solar energy may appear ideal and yet carry with it particular difficulties that become apparent later. Solar power has made enormous progress in the past half-decade, but five years is a short time to fully assess all the consequences of a new energy source.

In any endeavor that is changing as rapidly as solar energy, one must at some point decide to include certain subjects and exclude others. Since the time the bulk of the material included here was assembled, there has been growing interest in small-scale hydroelectric installations (sometimes called low-head hydro) and in satellite solar power stations that would transmit energy to earth via microwaves. Wave power is also a subject of renewed interest, but most of the research on this subject is being done in Britain and Japan.

Solar power systems in space would appear to face considerable economic difficulties, severe enough to put them into a separate category from earth-based solar systems. Low-head hydroelectric systems appear to have rather attractive economics, but the technology and the extent of the resource are uncertain. As with wave power, European research on low-head hydro systems is ahead of that in the United States.

For most of the major solar technologies, however, the United States is a world leader, and yet much of the best American work is known only to a limited audience. This book is an attempt to render as complete a description of American solar developments as possible, and to convey through specific examples—including over 90 illustrations—the wealth of activity occurring in the field of solar energy. My coauthor and I hope that you will find it a useful viewport for looking into the world of solar energy now and what it may be like in the future.

WILLIAM D. METZ

July 20, 1978
Washington, D.C.

SOLAR ENERGY IN AMERICA

1

STRATEGIES OF RESEARCH
Making Solar After the Nuclear Model

Solar energy is democratic. It falls on everyone and can be put to use by individuals and small groups of people. The public enthusiasm for solar is perhaps as much a reflection of its unusual accessibility as it is a vote for the environmental kindliness and inherent renewability of energy from the sun.

But the federal program to develop new energy technology is giving only belated recognition to solar energy's special characteristics. Despite the diffuse nature of the resource, the research program has emphasized large central stations to produce solar electricity in some distant future and has largely ignored small solar devices for producing on-site power—an approach one critic describes as "creating solar technologies in the image of nuclear power." The program contains virtually no significant projects to develop solar energy as a source of fuels and only modest efforts to exploit it as a source of heat. The massive engineering projects designed by aerospace companies which dominate much of the program seem to be intended for the existing utility industry—rather than individuals or communities—as the ultimate consumer of solar energy equipment.

One consequence of this R&D emphasis on large-scale, long-range systems is that it distorts economic and policy assessments of solar energy based on the current program, both within the federal energy research agency and in higher levels of the government. Indeed, the potential of solar energy is still regarded

1

with skepticism by many government energy officials and is publicly discounted by spokesmen for oil and electric utility companies. Funds for solar research are leveling off, because of the meager budget requests made by the Ford and Carter Administrations. Agency officials concede that even the present federal program—representing an investment amounting to less than one-half of that for new coal technologies and a small fraction of that committed to the nuclear field—has survived only because of the immense popular appeal of solar energy and consequent pressure from Congress.

In contrast to this official skepticism is the virtual explosion of optimism and activity elsewhere. Dozens of pieces of proposed solar legislation and hundreds of companies now manufacturing solar components reflect this interest. The number of solar-heated houses built in the United States has doubled approximately every 8 months since 1973, and the rate shows no sign of slackening. The rapid buildup of a fledgling industry has been matched or even exceeded by a staggering rate of technical innovation in designs for solar equipment and in research on advanced methods for capturing and using solar energy. Measured by the number of new ideas or the rate of progress, solar energy has become the hottest property and the most sought after action in the energy field. The burden of criticism from the solar energy community and from independent analysts is that the federal program has lagged rather than led many of these developments and that it has directed its research toward goals that betray a lack of understanding of the solar resource.

The government's difficulty in coming to grips with solar energy is understandable because the solar program was born, in an institutional sense, only about 6 years ago. The early work on solar energy was scattered among various government agencies, but much of it was an outgrowth of the effort by the National Aeronautics and Space Administration (NASA) to find practical spin-offs from space technology. After the 1969 Apollo moon landings, four different NASA laboratories began to do modest amounts of solar energy research. In 1972 the National Science Foundation became the lead agency for solar energy research, which was funded at only $2 million per year. Many of the early program managers came from NASA and much of the contracted research went to aerospace companies.

In early 1975, all the solar research programs were shifted from the National Science Foundation, which had traditionally been organized for basic research rather than commercial technology development, to the newly formed Energy Research and Development Administration (ERDA), where solar was cast into competition with the nuclear breeder, the government's reinvigorated coal program, and the growing program for fusion. In its first 2 years the ERDA solar program was greatly under staffed and the staff was overworked—at one time 60 percent of the mail for the entire agency concerned solar energy. But in spite of institutional handicaps, the program grew rapidly because Congress authorized large increases in the solar research budget—as much as 80 percent above what the agency officially requested.

The program under ERDA moved into a mode of design, construction, and testing of various types of solar power pilot plants on an aggressive timetable. Feeling pressure to build up the solar program rapidly, ERDA delegated a large—some critics would say dominant—role to its national laboratories and to various NASA laboratories. The different subprograms were evaluated in a series of "mission analysis" studies, largely performed by aerospace contractors, and new priorities were set. Much of the evaluation was based on the capability of various solar technologies to supply base-load electric power, under the assumption that anything else would fall short of a major contribution. During this crucial period of solidification, the program had no regular review by an outside advisory board and there were no congressional oversight hearings. One of the strongest outside influences on the shape of the program, according to well-informed observers, was the utility industry. Then, in the fall of 1977, ERDA was subsumed into the newly created Department of Energy, and the solar program was moved once more.

By 1977, government solar research had become a $290 million effort spread among four subprograms for electric applications, one for fuels, and two for heating, cooling, and related direct applications. The professional staff numbered about 70. In fiscal 1978, the program recommended by the Carter Administration grew only modestly to $320 million. For fiscal 1979, the administration budget was first reduced and then increased—because of the popular support evoked on "Sun Day," May 3, 1978—to $380 million.

Because the various solar technologies are generally unrelated to each other, there is not a great deal of overlap between the research bases needed for the subprograms. The result is that the different solar options are at an even greater disadvantage vis-à-vis other energy programs than the total solar research budget would indicate.

The largest allotment of funds and staff resources has been for solar electric technologies. The concept that the utility's research arm—the Electric Power Research Institute—sees as the most likely candidate for central electricity generation is the power tower (Figure 1), a system with a boiler on a high tower heated by the sunlight reflected from a field of hundreds or thousands of sun-following mirrors. The power tower with its related solar thermal systems is still the leading subprogram in dollar priority—it received 23 percent of the solar budget in fiscal 1979. Next is research on photovoltaic power systems—an effort to develop low-cost versions of the silicon cells used on space satellites for converting sunlight directly to electricity. Wind-power research, although it is the solar electric technology closest to being economically competitive, receives

only about 8 percent of the solar budget. Approximately 5 percent goes to develop methods for extracting energy from the small temperature differences between surface and deep seawater—a concept usually referred to as OTEC (ocean thermal energy conversion) and conceived to produce electricity or perhaps an energy-intensive chemical in a huge floating plant that would provide about 100 megawatts of power. Still less money goes to the solar resource that could be most versatile of all—plant matter or biomass, which can be converted into either heat, fuels, or electricity. Energy research officials are generally agreed that biomass is one area in which they have yet to get a strong and coherent program under way.

The solar heating and cooling subprogram has received about 30 percent of the solar research budget in recent years, and its portion is declining. Solar home and water heating (Figure 2) is nearly competitive in some areas of the country already. However, the successive energy agencies have been very slow to recognize the benefits of passive solar heating—the capture of solar heat that can be achieved by a well-sealed south-facing window, as opposed to a rooftop solar collector used with a water or airflow system to carry the heat downstairs. Such systems are now widely thought to be capable of filling a large fraction of the winter heating needs in many areas at costs generally less than those of active systems.

Solar energy is still in search of a proper institutional home. In 1977, Henry Marvin, solar program director under ERDA, said that the program had been subject to tight controls by the agency's upper echelons and by the Office of Management and Budget. In his words, "Congress has been the corrective factor" in the growth of the program, and budget data bear this out (Figure 3). Marvin is credited by several observers with having sought to limit the role in the program of the national laboratories—which, he says, "are not natural stopping places" en route to developing commercial technologies—and with having managed the program competently within the guidelines set by the agency. He has since been superseded as the director of solar energy research under the Department of Energy.

The program has been reorganized twice under the Department of Energy,

Figure 1. Power tower test facility at Albuquerque, New Mexico. Sunlight reflected off the mirrors in the foreground is focused onto a steel plate on the tower. Because of the large scale of the sun-tracking mirrors and the tower, the power tower concept requires immense amounts of steel and concrete—500 tons of steel and 2800 tons of concrete for every megawatt the plant produces. This is about 15 times the construction material needed by a nuclear plant and 35 times the amount needed by a coal plant. [Sandia Laboratories]

Figure 2. Water heating is one of the most practical solar applications. This house, in Boulder, Colorado, obtains hot water from the two solar collectors on the roof at the right. [Grumman Energy Systems, Inc.]

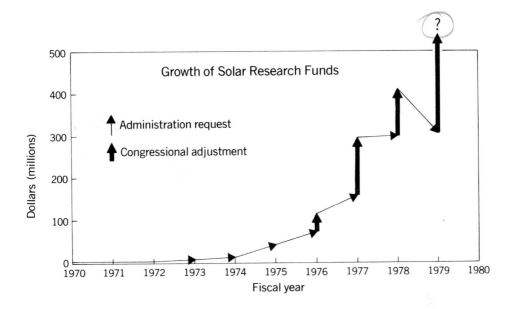

Figure 3. Annual budgets for solar energy research, as proposed by three administrations and revised—consistently upward—by Congress. The administration figures for 1978 and 1979 do not include supplements (of $20 million and $70 million) added midway through each budget cycle as the congressional intent became clear. [Adapted from data supplied by the Department of Energy]

but has undergone little effective change. It still largely reflects the narrow set of preconceptions with which it began. One of these preconceptions is the preferred role of centralized energy systems. Several pieces of evidence suggest that the ERDA program has given inadequate attention to the issue of the appropriate scale for solar technologies and in so doing, has failed to exploit the most promising characteristic of solar systems. A report by the congressional Office of Technology Assessment, for example, points out that federal research on electricity-generating equipment of all kinds has been focused almost exclusively on a centralized approach and has neglected a significant potential for on-site power production. The report—one of the most comprehensive studies of emerging solar technologies available—concludes that "devices having an output as small as a few kilowatts can be made as efficient as larger devices" and that on-site solar systems capable of generating electricity at prices competitive with those charged by utilities may be available "within 10 to 15 years." "On-site solar energy," the report declares, "must be regarded as an important option."

The solar thermal subprogram provides an instance of how ERDA's choices of scale were established. Initially, the subprogram was conceived of exclusively in terms of central power stations, as large as possible. Charles Grosskreutz, an analyst with the engineering firm of Black and Veatch during the period when it was involved in the initial program analysis of power towers for ERDA, said that "everyone started by considering a 1000 megawatt size and quickly scaled it down to about 100 megawatts" when it became clear that the tower height and the land acquisition problem were impractical for the larger sizes. "To my knowledge," he said, "there are no good studies of the optimum size of these facilities." Little serious consideration appears to have been given to solar thermal generating facilities in conjunction with community-scale energy systems or biomass fuel refineries—applications for which the optimum size, according to Princeton University physicists Robert Williams and Frank von Hippel, will probably be much less than 10 megawatts.

Likewise, the wind-power program, according to early program documents, did not look carefully at the prospects for improved versions of small wind turbines for distributed applications, or at the potential economies of mass production that might apply to small devices but not to large ones (Figures 4 and 5). Instead, the program plunged ahead to build a large, 100-kilowatt prototype as a first step toward a commercial size conceived to be as large as possible with the materials available—1.5 to 2.5 megawatts.

Williams and his colleagues point out that the solar program concentrates its main efforts on the largest and smallest scales of energy production, whereas an intermediate size may turn out to be the natural scale for many solar technologies. Their analysis points to community-size systems, equivalent to a few hundred or a few thousand houses, as the most cost-efficient, in that they would allow storage of solar energy on an annual basis—something impractical for an individual house—and would also allow the coproduction of solar heat and electricity in a manner that would be impractical for large central power plants.

Other independent analyses have come to similar conclusions. The noted British radio astronomer Martin Ryle, in a study of the applicability of solar energy to the United Kingdom, concluded that a distributed network of small wind turbines would provide the best match of potential supply to demand and would be competitive with coal-fired or nuclear generating stations. Ryle concluded that wind power, used with storage systems, could provide a substantial part of the power needs of the British Isles.

Another criticism of the solar program is that its management has been unnecessarily restrictive. During its period of rapid growth, the program has been guided by a management philosophy of "aggressive sequential" development. In practice, this has meant a policy of giving priority to one solar technology in each subprogram, such as the power tower in the solar thermal program, and pushing to quickly develop its hardware and test its feasibility. What the policy has ruled out—reportedly because of skepticism from the agency leadership and budget-cutting by the Office of Management and Budget—is the parallel development of competing concepts. It is, of course, possible that the best candidates were not chosen initially, but nevertheless a whole solar subprogram could be phased out because of poor performance by an ill-advised solar concept. In particular, features such as the scale and type of application have been heavily influenced by the original choices for development, and there is considerable danger that values derived from those choices will be the ones on which engineering and economic evaluations of future support will be made. It is just such considerations that lead enviromentalists to make the charge that solar energy is being "set up" to fail.

Another problem with the solar program has been lack of flexibility, leading to little integration of different solar technologies with each other and with the energy needs they might ultimately satisfy. Storage is a problem with many solar systems, but the program has given little attention to applications in which biomass fuels would provide the storage element, or in which the need for storage would be obviated by using solar energy in conjunction with another energy source. Solar-coal and solar-hydroelectric systems offer tantalizing possibilities for combinations that could approach around-the-clock power, and there is some evidence that direct solar energy and wind energy might complement each other well. Little attention has been given to on-site application of photovoltaic and solar thermal devices, in which the utility grid could be used as a buffer and thus storage would not be required. In addition, a generally acknowledged problem with the ERDA program was that its sharply divided subprogram structure has limited the development of systems that serve two purposes at once, such as "total energy" systems that produce both heat and electricity with a considerable improvement over the efficiency of single-purpose systems (Figure 6). The program only belatedly began to look at projects that did not fall into any of the predefined categories, such as solar irrigation, which ERDA developed no sooner than did the state of Guanajuato, Mexico.

The organizational structure of ERDA, moreover, was at cross-purposes with many novel or noncentralized applications. The solar energy division, for example, was effectively prohibited from working on community-scale solar systems because the agency management decreed community-oriented projects to be in the domain of the conservation directorate. The reorganization of the program under the Department of Energy carries the risk of even greater fragmentation.

Cost is the stumbling block most often cited by solar skeptics, and there is no doubt that few of the solar options are competitive today. But in the absence of a real market, current cost estimates are almost certainly deceptive. Furthermore, no one really knows what the costs of small-scale systems will be because so little research has been done on them. The conventional wisdom at the solar program planning office is that, compared to electricity at current prices, wind generators are competitive today or within a factor of 2 of being competitive,

Figures 4 and 5. Small wind systems are now competitive with large systems and may soon prove cheaper. The federal wind research program has been predominantly devoted to large machines. Two wind turbines of modern design are shown here. The large, 200-kilowatt machine (facing page) was developed by the Department of Energy and installed at Clayton, New Mexico, where it started operation in early 1978. The small, 15-kilowatt machine (above) was privately developed by Grumman and installed on a dairy farm in Hopkinton, New York, where it started operation in mid-1977. [Department of Energy and Grumman Energy Systems, Inc.]

Figure 6. The diversity and flexibility of solar components are reflected in the number of ways they can be put together. Here concentrating lenses made of plastic (right) focus sunlight onto an array of photovoltaic cells (bright spots on the left). The concentrating photovoltaic system produces electricity at enhanced efficiency and also hot water. [Sandia Laboratories]

biomass fuels are a factor of 2 to 4 away from a competitive price, ocean thermal power systems a factor of 4 to 5, power towers a factor of 5 to 10, and photovoltaics a factor of 10 to 20 away. The opportunities for price reduction among these different technologies are controlled by quite different factors, however. Technologies as diverse as passive heating, solar-assisted waste treatment, and concentrating collectors (Figures 7 to 9) will be built by groups with business styles as different as those of small builders, large municipal construction companies, and nationwide manufacturing organizations. While collectors are clearly mass-producible, passive houses almost certainly are not. Even the technologies for which a market does exist—water heating, for example—do not yet benefit from the kinds of implicit subsidies enjoyed by most other energy sources or the advantages of mass production by a well-established industry.

Probably no question about solar is more controversial than whether it can become a major energy source in the near term or should be regarded (and funded) as a limited, long-range option. Assessments of this question tend to be swept up into what has become a highly polarized debate between environmental advocates and the defenders of coal and nuclear power—a debate whose terms are more nearly philosophical or ethical than economic. The one view holds that a transition to a predominantly solar economy is not only feasible but *neces-*

Figure 7. The Karen Terry house, an example of south-facing design in a passively heated and cooled solar house, Santa Fe, New Mexico. [Copyright © New Mexico Solar Energy Association 1978. All rights reserved.]

Figure 8.   Solar collectors used for a municipal waste treatment plant. The flat-plate collectors supply heat to sludge digesters, enabling the plant, in Wilton, Maine, to produce methane gas. [Grumman Energy Systems, Inc.]

Figure 9. A trough-like solar collector concentrates sunlight up to 50 times its normal intensity. The curved mirror focuses sunlight onto the evacuated tube above and follows the sun by tilting upward as the sun moves higher in the sky. Such one-dimensional tracking collectors produce higher temperatures and are considerably more efficient than stationary, flat collectors. [Honeywell Energy Resources Center]

sary—to avert climatic disaster from the buildup of carbon dioxide that would accompany massive use of coal, and to prevent the danger of nuclear warfare attendant on the proliferation of the plutonium economy. The other dismisses solar energy and holds that coal and nuclear are essential, on the grounds that even if costs were to drop dramatically it would still be many decades before enough solar-heated houses and solar power stations could be built to make any dent in this country's huge and growing appetite for energy.

But these tactical positions obscure a number of things that tend to argue the importance of solar energy on purely economic grounds, as well as obscuring some substantial problems. One of the key problems is that solar equipment tends to be capital-intensive, with high initial costs that are a deterrent to consumers unaccustomed to making decisions on a life-cycle basis. Another is that many existing institutional arrangements, from building codes to utility rate structures to federal tax policies, discriminate against unconventional energy sources. But some institutional barriers are being removed by legislation, and the prices of many solar components are already dropping sharply in response to steadily growing demand. It seems evident that the growth of distributed solar systems, for which equipment can be mass-produced, can be far more rapid than the growth of centralized power plants, which must be laboriously assembled in the field. Frost and Sullivan, a respected market research firm, predicts that 2.5 million U.S. homes will be solar heated by 1985. The government itself may become a major market for solar energy—a Department of Defense report prepared for the Federal Energy Administration (now part of the Department of Energy) estimated that there might be a defense market for up to 100 megawatts worth of photovoltaic devices a year at the prices expected to prevail in the early 1980s.

Political fortunes may also play a role in determining the short- or long-term impact. Solar energy fared badly under Republican administrations. President Ford had many opportunities to attend solar project ribbon cuttings but did not do so. Under his administration, the Office of Management and Budget strenuously opposed and nearly gutted the major short-term elements of the government's solar energy program—the demonstration projects for solar heating. ERDA appealed to President Ford but, according to one observer, had the misfortune to argue its case during a week in which Ford was preoccupied with the Angolan crisis. In any case, the Office of Management and Budget position largely prevailed—a circumstance that apparently contributed substantially to the resignation of ERDA assistant administrator John Teem—and the proposed demonstration program, modest though it was, was drastically cut back.

The government program has had some effect—ERDA's work on photovoltaics and wind stimulated some private investment. And quite apart from the government's program, there appears to be a remarkable amount of momentum in solar thermal devices, wood-burning stoves and boilers, and other components of a solar energy industry.

After 6 years of rapid but uneven development, solar energy is in need of reassessment. The present federal program has been as much the product of institutional happenstance and various technical predilections as it has been the product of coherent planning. In a broader perspective, the government policy has been to characterize solar energy as a long-term option comparable to fusion and the breeder reactor, but in fact it has little in common with these potential leviathans. Solar technology is more diverse, and even the most difficult technologies, such as photovoltaics, may be closer to commercial realization.

Many solar technologies already work and they are already facing the economic challenges that other long-range options have yet to confront. As subsequent chapters will illustrate, the progress in solar research has been dramatic. It is arguably time to reconsider priorities and ask whether the distribution of research resources among nuclear, fossil, and solar options reflects a rational policy.

2

STEEL AND CONCRETE
Power Tower Dominates Solar Research

The centerpiece of the government's solar energy research program is proceeding in a fashion that is remarkably parallel to the pattern of nuclear power development. Limited focus on a narrow technical option, a large investment in test facilities, increasing expenditures for a series of prototype power stations, and a multidecade schedule all characterize the U.S. program for developing centralized solar thermal generating stations.

The eventual cost may or may not be as large as the bill for developing a new reactor system, but the yearly payment is already great enough to dominate the budget of the relatively small solar program. The project to develop the power tower receives about one-fourth of the solar energy research support. Because the power tower program has jumped quickly into the construction phase, it has garnered more of the solar budget than either of the programs for developing photovoltaic cells or wind power—alternative ways to produce electricity through solar energy. Yet at about $10,000 per peak watt, the present power tower costs are at least as high as those of photocells and markedly higher than those of wind systems.

The concept is the brainchild of the aerospace industry, which conducted the studies that the Energy Research and Development Administration used in deciding how to set up its solar thermal electric program. Since the program has been under way, most of the research has been contracted out to aerospace companies. The approach of the energy agency was to obtain as many as four designs for each component of the system and to evaluate those designs by use of a prototype. Inherited by ERDA's successor, the Department of Energy, the program has been given a vote of confidence and continues to expand. It is still far too early to judge how successful the power tower will be, but it is the paramount example of the tendency of the U.S. solar energy program to favor centralized solar concepts at the expense of other options.

A power tower is a system for collecting solar energy from a large field of mirrors and converting it into heat at high temperatures for efficient generation of electricity. Optical studies show that the best way to get the high temperatures is with a point-focusing mirror that tracks the sun (a heliostat), and systems studies made by the Aerospace Corporation in 1974 found that the cheapest way to combine the heat from many such mirrors is to focus them all on a single boiler set atop a large tower. Heliostats that concentrate sunlight 1000-fold (a concentration ratio of 1000) are used to raise the temperature in the boiler to 500°C, and steam from the boiler can produce electricity in a conventional turbogenerator.

Some engineers favor a central system using mirrors that are not point-focusing, and others question whether the cost savings are not offset by practical problems—the tower for a 100-megawatt plant would be about 1000 feet tall. But both ERDA and the Electric Power Research Institute are betting heavily that the power tower, with the benefits of "photon energy transport," is the best design for a centralized solar thermal generating plant.

Although alternative centralized generating concepts suffer reduced support in comparison to the power tower, the research area that appears to be hardest hit by the power tower's generous funding is the development of intermediate-temperature solar thermal systems that could be used on a smaller scale and of distributed electric systems. Systems that employ mirrors that are less optically sophisticated than those of the power tower can convert sunlight to heat in a very useful temperature range, above the 80°C limit of flat rooftop collectors and below the 500°C level achieved by point-focusing heliostats. The intermediate performance mirrors are generally variations of parabolic troughs—line-focusing elements that track the sun in only one direction during the day. They concentrate sunlight by a factor of 10 to 50 and focus it onto evacuated tubes suspended above the troughs.

Intermediate-temperature systems are slightly less efficient than heliostats if used for electricity generation alone, but they are simpler, cheaper, and more readily adapted to applications where, in addition to electricity, they produce heat for warming or cooling. A report prepared by the Office of Technology Assessment in 1977 estimated that a parabolic trough collector would have a useful annual output (that is, energy delivered after thermal and optical losses are taken into account) only 10 percent less than that of a heliostat. Because of the lower temperature, a parabolic trough system would convert less energy into electricity. But, by utilizing the waste heat, such systems could achieve an overall efficiency that would exceed the typical efficiency (16 percent) expected for a power tower central station. Intermediate-temperature collectors and systems that use them to produce both heat and electricity will be discussed further in Chapters 4 and 5.

The economics of small systems and intermediate-temperature systems are not well known—in large part because so little money has been available to study them. But one of the most striking conclusions of the Office of Technology Assessment report was that there is "no clear indication that large solar electric plants are more efficient or produce less costly energy than small, on-site facilities."

Recent changes have upgraded research on solar electric systems for nonutility applications, but the bulk of solar electric research is devoted to technologies designed exclusively for large electric utilities. In the solar thermal subprogram, ERDA spent $60 million on central systems in fiscal 1977 (almost all of it for the power tower), while allocating $9 million to total energy systems. This was despite the fact that small systems have the potential for making energy contributions in the near future. Very large solar thermal electric stations, because they are being developed by the same process as nuclear stations, are unlikely to make a contribution to commercial energy supplies in less than 20 years.

Thus, much like the breeder reactor, the power tower is scheduled to proceed from a small test plant to the first commercial plant in four stages. If bar graphs showing the scheduled development stages for the two projects were superimposed on each other, they would show a striking similarity. The first solar stage is a 5-megawatt thermal (MWt) test facility that has just been completed at Sandia Laboratories near Albuquerque, New Mexico, for $21 million (Figure 10). The second will be a 10-megawatt electric (MWe) plant to be built near Barstow, California, at a cost of $130 million; construction is due to begin in the fall of 1978. These two stages, funded largely by the government, are due to be followed by a 100-MWe demonstration plant in the mid-1980s, and finally a 100-MWe prototype commercial plant in the 1990s. As with the breeder, the government hopes to share major parts of the costs of the latter two projects with the utilities that will use them. One difference in the solar case is that the utilities, through the Electric Power Research Institute, are contributing small amounts of funding for studies in parallel with the first phase of the government program, and a California utility group is contributing $20 million to the Barstow project. The utility group is comprised of Southern California Edison, the Los Angeles Department of Water and Power, and the California Energy Commission.

Figure 10. Aerial view of the $21 million power tower test facility at Albuquerque, New Mexico. Seventy-eight sun-tracking mirrors (heliostats) have been installed, and eventually 200 more will be added. The mirrors focus sunlight on the top of the 200-foot tower, on which experimental boilers (or receivers) will be tested. The pickup trucks in the background indicate the scale of the facility. On completion, this test facility will have a peak production of 5 megawatts of thermal power. Because of the power lost in converting thermal energy to electricity, the planned 10-megawatt electric pilot plant at Barstow, California, will have to be 12 times larger than the Albuquerque test facility. A commercial power tower plant (now projected at 100 megawatts electric) would have to be 10 times larger again. Under the concentrations of light produced by a power tower field, steel can melt or deform. One advantage of concrete construction is its superior performance under accidental exposure to concentrated sunlight. [Sandia Laboratories]

Whereas the government's nuclear program nurtured four large heavy-equipment companies that are now the sole suppliers of nuclear reactors in the United States, the power tower program is dispensing the bulk of its work to four large aerospace contractors. The companies that have built test hardware for the Albuquerque facility and are competing for contracts on the much larger Barstow facility are Martin Marietta, Honeywell, McDonnell Douglas, and Boeing. If the power tower proceeds apace, their names will become as synonymous with solar electricity as the names Westinghouse, General Electric, Combustion Engineering, and Babcock and Wilcox have become with nuclear power. The success of the power tower concept will probably hinge on the development of the novel high-technology components—the collectors, receivers, and thermal storage units. But the rate of development will more likely be controlled by the logistics of designing and building the sequence of solar plants, each of which will be a huge conventional construction project.

Commenting on the apparent similarities with nuclear development, the head of the Department of Energy's solar thermal branch, Gerald Braun, says that common features are less deliberate than automatic. "Because you are looking at something at the same scale," says Braun, "you go the same way. But there was certainly no intention to follow the nuclear model."

The name power tower has a friendly ring to it which conjures up something on a human scale. But an aerial view of the Albuquerque facility (Figure 10) indicates how large a full-scale system will be. Even though it will have only one-twelfth the thermal capacity of the Barstow plant, the Albuquerque tower is as high as a 20-story building and is built with 5700 cubic yards of concrete. The collectors at the Albuquerque test site will cover 100 acres. Each heliostat is anchored with a 10-ton concrete footing (Figure 11). A plant with greater power will require a larger collector field and a higher tower. The tower for the Barstow plant will be about 500 feet high, and double that height will be needed for the customarily projected commercial-size plant producing 100 megawatts.

A commercial power tower generating plant would cover about 1 square mile of land, probably at a desert site, collecting sunlight from as many as 10,000 heliostats. In the designs of the four contractors, the heliostats are 37 square meters in area. To counter the effect of passing clouds, the Barstow pilot plant will have a thermal storage capability for 3 hours of electricity generation—the storage system will be a huge tank filled with oil and rock. Although the storage could allow operation to continue briefly into the early evening, the plants are primarily intended to supply electricity to meet the midday load (often called an intermediate load) experienced by most utilities. The Department of Energy plans call for a steam (Rankine) cycle that would require a considerable amount of cooling water, but the concept favored by the Electric Power Research Institute (a Brayton or gas turbine cycle) would require little or no cooling (Figure 12).

Figure 11. Individual heliostats at the Albuquerque test facility. Each assembly can move in two directions to follow the sun across the sky at all times of the year. Each heliostat is made up of 25 4-foot-square mirrors. On a clear day in Albuquerque, the amount of sunlight falling on a properly aligned heliostat approaches 1 kilowatt per square meter. The heliostat is turned face down to protect the mirrors during storms. [Sandia Laboratories]

The Department of Energy power tower would be cooled by evaporation and would use at least as much water as a fossil plant of comparable size.

According to most estimates, the major factor influencing the costs of the power tower plant will be the design and cost of the collector. The shape of each collector must approximate a parabola focused on the receiver, but the four aerospace contractors recommended rather different ways to accomplish that end. Three of the contractors designed mirrors to be made of steel and glass, while the fourth (Boeing) designed a mirror to be made of aluminized polyester stretched across a circular frame (Figure 13). The Honeywell design had rectangular mirrors mounted on a geared tracking frame that tilts in two directions. The design by McDonnell Douglas used a solid dish, made of eight mirror segments, mounted on a radar pedestal. The Martin Marietta heliostat used nine mirrors mounted on a common tracking frame. In some cases, the flat mirrors had to be stressed slightly to give a parabolic shape, and the facets had to be aligned to point at a common focus. In late 1977, the overall design proposed by McDonnell Douglas was chosen for Barstow.

The Boeing heliostat is potentially less expensive than the others because it uses very lightweight materials (Figure 14). The collector is protected from weather, wind, and dust by a plastic bubble supported by air pressure. Boeing estimates that the optical loss that occurs when sunlight passes through the bubble (about 20 percent) is more than compensated for by the cost reduction achieved by using light, thin materials. All four heliostats follow the sun by tracking along two axes. The Boeing and Honeywell versions are directed by computer control. The other two versions are controlled by a feedback signal from a sensor in the reflected beam of each heliostat.

Initially, the four contractors each built a half-dozen heliostats. Their costs were quite high—in the range of $500 to $1000 per square meter. One critic commented that, so far, the heliostats being produced by the aerospace firms are being delivered at "aerospace prices." This is also the price range of satellite-tracking antennas—parabolic dishes that resemble heliostats in a number of ways. Clearly, an important challenge to the power tower program is the problem of reducing these costs. The official goal of the Department of Energy program, which many observers consider unattainable, is a cost of $70 per square meter.

Heliostat cost reduction is crucial because the heliostats represent about 60 percent of the total cost of a power tower plant. There has been a wide range in early cost estimates prepared by McDonnell Douglas, Martin Marietta, and Honeywell—from $40 to $96 per square meter. Because their durable lifetime is well known, glass and steel heliostats were made a requirement for the Barstow plant. In a study of the detailed economics of many types of solar electric systems, Richard Caputo at the Jet Propulsion Laboratory concluded that with careful

development work a glass and steel heliostat cost of $145 per square meter (in 1975 dollars) should be attainable. But he indicated that the costs could go higher. At a cost of $145 per square meter, he calculated that the plant capital cost would be $2000 per kilowatt electric. The Boeing heliostat should be producible at $50 per square meter in quantities of 25,000 per year, according to Roger Gilette at Boeing. The economics of the Boeing design could be unfavorable if it has a shorter lifetime, however, and its useful lifetime under working conditions is still to be determined.

Another important factor is the design of the receiver that will sit atop the tower and absorb the concentrated sunlight. Two of the competing contractors, Martin Marietta and Honeywell, were planning to use cavity-type receivers, while the third, McDonnell Douglas, proposed an external receiver. The cavity type would have been more limited in the total power that one receiver could absorb, particularly in the Martin Marietta design. The McDonnell Douglas receiver, which has heat exchanger tubes on the outside of the tower, can be more easily scaled to a large plant and was the one chosen by the energy agency.

Many engineers think that the development of a reliable receiver will be a most difficult task. Since the power tower produces higher concentrations of sunlight than occur in other collectors, the receiver on a power tower must endure

Figure 12. Artist's rendition of a 60-megawatt solar thermal generating plant, following the design of an Electric Power Research Institute study. Sunlight would be delivered to the central tower by 7000 heliostats covering one-quarter square mile. The central tower, 700 feet tall, would have a receiver for the focused sunlight on the highest level and a gas turbine generator on a lower level. Hot (1038°C) air from the receiver would produce power in the turbine and then be exhausted. The open-cycle gas turbine system would achieve a slightly higher efficiency than a closed-cycle steam system, and no cooling towers would be needed. Rather than use a heat storage tank for backup—at an estimated $180 per kilowatt of total capacity—the design would use fossil fuel to fire the gas turbine at a cost of $4 per kilowatt for extra equipment. The cost of the plant would be $1250 to $1660 per kilowatt (1976 dollars), based on collector costs in the range of $60 to $125 per square meter. [Electric Power Research Institute]

considerably more difficult conditions. The materials used in the boiler must be able to withstand instantaneous changes in energy density from 0 to 5 megawatts per square meter. Although the receiver may be kept operating overnight at a reduced temperature to ease the problem of starting it up each morning, the system will be subject to frequent temperature cycles. According to Charles Backus at Arizona State University in Tempe, "The design of the absorber or boiler for this concept will be the major technical challenge."

The analysis contained in a power plant study conducted for the Electric Power Research Institute tends to substantiate and—if anything—carry further Backus' critique of the problems to be encountered in developing a power tower receiver. EPRI's study, under contract to Black and Veatch, of a power tower with a gas turbine generator found that the best operating temperature, twice that of a steam turbine system, would be so high that it would rule out the use of metals in the high-temperature face of the receiver cavity. Instead, it recommended ceramic (silicon carbide) heat exchanger tubes and highlighted problems with installing the ceramic tubes and suitably insulating the hot-air ducts in the plant.

The gas turbine power tower study also took a novel approach to the problem of energy storage. Rather than use a huge and expensive tank for energy storage,

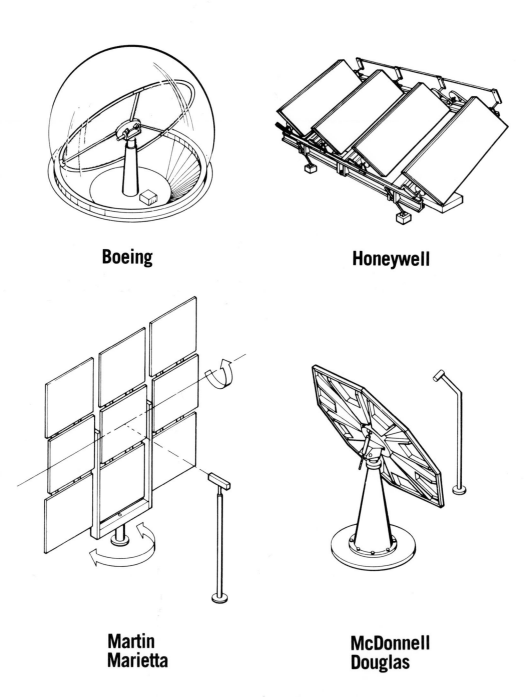

Boeing

Honeywell

Martin Marietta

McDonnell Douglas

Figure 13. Four different heliostat designs originally proposed for the Department of Energy's 10-megawatt solar thermal pilot plant. According to the government specifications, each has an area of at least 37 square meters. [Office of Technology Assessment]

as the Department of Energy concept would, it would use fossil fuel (oil or gas) firing of the solar turbine as a backup option, one that would add very little extra cost. The tower would have two large decks at the top to hold the 750-ton receiver and the 880-ton turbine. (The EPRI study found that providing stability against wind and seismic activity for the heavy load at the top would be the principal structural problem in designing the tower.) With a higher operating temperature, the overall efficiency of the system was calculated as 18 percent and the cost was estimated between $1250 and $1660 per kilowatt, depending on heliostat costs. The study concluded that hybrid (fossil-solar) operation was feasible and in fact desirable, since fossil fuel can serve the purposes of both short- and long-term energy storage.

Although most power tower research is focused on large towers and large heliostats, there is a smaller, cheaper, simpler power tower that has been operating successfully for more than a decade. First tested by Giovanni Francia at the University of Genoa in 1965, the system has mirrors 1 meter in size, controlled by a common mechanical drive, and collects the light in a receiver hung from a short, lightweight steel tower. The system is now available in a package from AN-SALDO, SpA, the major heavy electrical equipment manufacturer of Italy. A 400-kilowatt thermal (kWt) system, built by ANSALDO and delivered to the Georgia Institute of Technology, is now operating (Figure 15).

With a range from 400-kWt working models to 60- to 150-MWe conceptual studies, the optimum size of a power tower plant is a matter of much guessing and some reevaluation just now. There is some indication that solar administrators may rethink the validity of the 100-MWe goal chosen 4 years ago. Charles Grosskreutz, now at the newly inaugurated Solar Energy Research Institute, says that the 100-MWe size was not "the result of a careful study in which someone found a curve with a dip in it," but a reaction to practical limiting factors. The administrator who was in charge of evaluating the competing design studies, Alan Skinrood at Sandia Laboratories in Livermore, California, says that it is "fairly clear" that the optimum plant size for the United States is 50 to 200 MWe.

Whatever the outcome of the aerospace competition, it is clear that the costs of power tower systems are far too high. The present price of collectors is about ten times the $70 goal, and the total pilot plant cost—which will be over $10,000 per kilowatt for the Barstow facility—are in excess of those in other energy technologies that have reached a similar stage of development. The plan of the solar research program is to reduce these costs—particularly for heliostats—by bulk manufacturing techniques and steadily improved designs.

The high cost of the power tower would be less troublesome if there were equally high prospects for cost reduction. But if the various solar technologies were ranked according to their cost breakthrough potential, the power tower would have to be near the bottom. The reason for this is that the power tower

Figure 14. Lightweight, low-cost helio-stat designed by Boeing for consideration in power tower development. The circular reflector is protected from the elements by an air-supported plastic bubble 17 feet in diameter. The reflector, made of alumin-ized Mylar, is 15 feet in diameter. The Mylar is stretched to the proper tension to form a concentrating paraboloid under its own weight. This test unit, located near Boardman, Oregon, has withstood the highest winds that have occurred so far at the site—up to 70 miles per hour. The bubble enclosure, made of 4-mil-thick Ted-lar, is designed to withstand winds of 90 miles per hour. The reflector is set in a gimbal made of tubular aluminum and steered by computer control with a small stepping motor. The entire unit, excluding the steel ring wall and the concrete foun-dation, weighs about 200 pounds. The unit will reflect 70 percent of the incident sun-light onto a tower. Boeing submitted a bid to produce this design for $300 per square meter of collector area for the Barstow project, if 1500 collectors were built.

The company has designed an im-proved version of the collector that will be 32 feet in diameter. The base will be made lighter by substituting a steel pipe ring on three legs for the ring wall, to save on materials costs. The bubble will be made of a cheaper material—weatherized poly-ester—which can be more easily mass-pro-duced in a dome shape by thermal forming into a large free-blown bubble. Boeing es-timates that the new design can be made for less than $50 per square meter in quan-tity. [Boeing Engineering and Construc-tion]

requires huge amounts of steel and concrete—about 15 times the construction material needed by a nuclear plant and 35 times the amount needed by a coal plant. One must be very skeptical that such a material-intensive technology can achieve the tenfold cost reduction needed to compete with wind and other solar electric systems. A study recently done for the Department of Energy by SRI International concluded that the power tower would not compete economically for the next 40 years.

Other types of solar thermal technology besides the power tower may be much more attractive. These include distributed electric systems, total energy systems, and systems for the industrial utilization of solar energy to produce process heat.

But the benefits of systems on smaller scales may be forgotten by the time the final answer for the power tower is known. In particular, the rule of thumb that energy can be transported more economically by light than by heat may be true only for large systems. Heat losses in piping depend on the average distance in the heat transport system, so the economics of distributed collectors connected to a generator by heat (rather than light) transport should become favorable at some point for smaller systems. For much the same reason, total energy systems should be preferable on small scales. But the energy agency appears to be

supporting such projects principally as a backup in case the power tower project should fail.

The history of the nuclear development program offers some lessons in the danger of overconcentration of effort on one or two technical concepts. In a huge development program, the ideas of talented workers may be wasted because of the necessity of working within rigid management structures on programs with externally imposed goals. For a number of technologies there is very little choice. But for solar energy, even for the specific purpose of converting it to electricity via thermal systems, there are many choices, and new inventions are appearing rapidly. It would appear to be far too soon for the solar program to be discarding innovative options and sinking its research money into steel and concrete.

$$3$$

PHOTOVOLTAIC CELLS
The Semiconductor Revolution Comes to Solar

If there is a dream solar technology it is probably photovoltaics—solar cells. These devices convert sunlight to electricity directly, bypassing thermodynamic cycles and mechanical generators altogether. They have no moving parts and are consequently quiet, extremely reliable, and easy to operate. Photovoltaic cells are a space age electronic marvel, at once the most sophisticated solar technology and the simplest, the most environmentally benign source of electricity yet conceived.

At present the only cells available commercially are made of silicon crystals and assembled largely by hand, an expensive proposition. But nearly two dozen additional types of photovoltaic cells and a variety of automated production techniques are under development. In fact, photovoltaic technology is now undergoing a burst of innovation comparable in many ways to that which gave rise to another semiconductor technology, integrated circuits. New and more efficient cell designs have been proposed, capable of converting between 30 and 40 percent of the sunlight falling on them to electricity. Other novel designs make use of inexpensive starting materials or lend themselves to automated production. Experiments with new fabrication techniques ranging from spraying or depositing thin films of photovoltaic material to continuous production of silicon crystal are underway. The proliferation of photovoltaic technologies has been so rapid as to outdate assessments of the field made just a few years ago. It has convinced many photovoltaic researchers and analysts that solar cells can become a significant source of energy by the end of the century.

The federal energy program, however, has not sought aggressively to realize the photovoltaic dream. Despite growing evidence that such devices could become a competitive source of on-site power within a decade, the Department of Energy does not regard photovoltaics as a significant energy option for this century. Disregarding both the ambitious goals of its own research program and the striking progress achieved to date, the agency allotted only $76 million to photovoltaics in the fiscal year 1979 budget—less than one-fourth of the money being spent on the 21st century technology of nuclear fusion.

Photovoltaics are a practical source of power now for remote applications, with commercial sales of about 1 megawatt of generating capacity expected in 1978 (Figure 16). Prices are decreasing rapidly at a rate exceeding 30 percent per year, but they are still high, about $10 per peak watt for flat-plate arrays and $6 per peak watt for concentrating arrays. Prices must drop 10- to 20-fold before photovoltaic cells could come into general use as a source of on-site power, and further reductions are needed before central power stations would be feasible.

Price reductions of a factor of 10 or more are common in the semiconductor industry for products with large markets, as witness the history of hand calculators and digital watches. But they are nearly unheard of in the energy industry. Federal energy planners and other officials whose experience is primarily with conventional thermal or nuclear energy systems are unfamiliar with the manu-facturing processes applicable to photovoltaic cells and related semiconductor devices, and they find the possibility of such staggering cost reductions hard to believe. Their skepticism, widely shared within the utility industry, has been translated into tight R&D budgets, a reluctance to stimulate production through federal purchases, and a general lack of support of what could be the most revolutionary solar technology of all.

Ironically, the federal photovoltaic research effort is credited by many observers with being perhaps the best conceived and most successful of the government solar programs. Despite its designation as a long-term option, the program has an ambitious set of goals that would make photovoltaic power widely available by 1986. Not only is it achieving improvements in the efficiency and reductions in the cost of silicon solar cells at a more rapid rate than that projected by its plan, but it also appears to have stimulated private industry into activity. Researchers at several major semiconductor manufacturers say that the Department of Energy program has attracted the attention of corporate manage-ment to the potential for near-term markets and resulted in the establishment or upgrading of proprietary development efforts. Private efforts have accelerated in recent years, and there are even indications that the smart money is betting on

Figure 16. Photovoltaic arrays at a remote site in the Mohave desert. These arrays contain more than 2000 individual photovoltaic cells and can generate 460 watts in peak sunshine. The electricity is used to power a repeater station for microwave and ultrahigh-frequency television signals. In the background are storage tanks for a propane generator that proved unreliable. Apart from federally sponsored demonstration projects, most photovoltaic power systems are now sold for remote installations such as this one. Another market being eyed with interest by photovoltaic cell manufacturers is villages in developing countries, many of which are also remote from modern electric power grids and lack electricity for lighting, refrigeration, and communication systems. A particular advantage of photovoltaic systems for remote applications of any kind is that they require little if any maintenance and do not require fuel.

[Solarex Corporation]

photovoltaics—oil companies, including at least four major firms, have concentrated their solar investments in photovoltaic technologies. The second largest producer of solar cells, for example, is a subsidiary of Exxon.

Cells made from large crystals of silicon are the dominant commercial technology. They were developed at Bell Laboratories in the early 1950s, not long after a related semiconductor device, the transistor. Unlike the situation with transistors, however, there was at the time no prospect of a mass market. Instead, silicon cells were developed for space applications and the emphasis was on reducing weight, not cost. Even now, cells manufactured for terrestrial applications (and sold for prices 50 times lower than those that prevailed a few years ago in the space market) require between 10 and 100 man-hours of handwork per kilowatt of generating capacity. Complete automation has yet to be achieved.

For silicon cells, cost reductions may depend more on the application of mass production methods to known techniques than on fundamental breakthroughs or new concepts. That at least has been the thrust of the federal photovoltaic R&D program, which has been focused largely on silicon cells, and that is the assumption on which predictions of commercially competitive photovoltaic power by the late 1980s are usually based. A concomitant part of the strategy is to build up a photovoltaic industry through large federal purchases of cells—stimulating production and lowering costs by helping to create an interim market.

Much of the optimism regarding low-cost silicon cells stems from recent systems studies and laboratory work done for the Jet Propulsion Laboratory, which is managing the flat-plate silicon program for the Department of Energy. Although the effort has so far generated more paper than cells, it is credited with having tapped some of the best industrial talent in the country. One key area is material costs—the extremely pure semiconductor grade of silicon now used to make cells costs $65 per kilogram and is a substantial component in the cost of the final product. Moreover, cells made with present techniques would have to operate about 12 years to recoup the energy expended in their manufacture, in part because the raw silicon is melted and remelted so many times during purification and crystal growth that it is one of the most energy-intensive commercial materials in the world. But studies by Dow Corning and Union Carbide, among others, indicate that a sixfold decrease in the cost and a tenfold or greater reduction in the manufacturing energy for solar-grade silicon are feasible. Another major problem with present techniques is waste. About 80 percent of the purified silicon used to grow large cylinders of the material in crystalline form is left as scrap in the crucibles, or ends up as sawdust when the thin wafers used to make cells are cut from the cylinder. Methods of growing larger cylinders, of recharging the crucibles, and of sawing thinner slices—for example, by a new laser slicing technique being developed by Texas Instruments—are expected to reduce this waste by at least half.

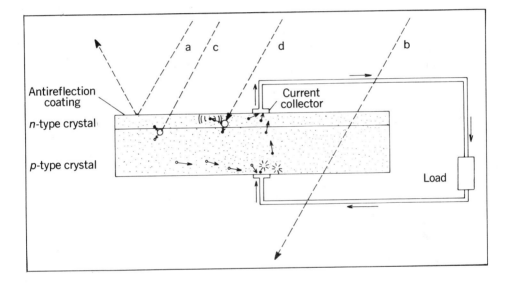

Figure 17. A silicon solar cell is a relatively simple semiconductor device compared to modern integrated circuits such as those used in hand calculators. The essential structure is shown here. Light hitting the cell may be reflected (a) and thus lost; special coatings designed to reduce reflection are applied to most commercial cells. Light may also pass completely through the cell (b) if its wavelength is too long. Light that is absorbed within a cell (c and d) interacts with a silicon atom to produce a free electron bearing negative charge and its opposite, a positively charged vacancy or hole in the crystal's electronic structure. Impurities introduced into the silicon crystal when the cell is manufactured create a semiconductor junction—the boundary between n-type and p-type silicon—which generates an internal electric field within the cell. Free electrons are induced by the field to migrate to the top of the cell, while positive charges migrate to the bottom, thus creating a flow of current whenever light is absorbed by the cell. The current is collected by a grid of contacts on the top and bottom of the cell. Most photovoltaic cells operate in a similar way, but use different types of semiconductor junctions. [Research Triangle Institute]

The bulk of the cost of a silicon cell, however, comes from the many mechanical steps required to convert the raw wafer into a commercially useful product. Controlled amounts of impurities are diffused into the silicon to create a *p-n* junction—which creates a kind of internal electric field that propels positive charges in one direction and negative charges in the other (Figure 17). (Pairs of such charges are created when sunlight is absorbed in the cell.) A grid of metal contacts must be attached to the front and rear of the cell to collect the charges that migrate there and thus create a flow of current. Finally, individual cells must be assembled into an array and encapsulated to protect against deterioration.

Much of this is now done by hand, but studies by RCA, Texas Instruments, and Motorola have indicated that the cell manufacture and assembly steps can be streamlined and largely automated with substantial reductions in cost. Beyond that, improvements that are well established but not now used in solar cell manufacture are being considered—such as the use of ion implantation techniques for introducing impurities during cell manufacture. Apparently such studies have convinced these companies, which are experienced in assessing and manufacturing semiconductor products, that silicon cells have a future. Motorola recently announced that it is entering the solar cell business, and the others are known to be studying the prospect closely.

Still further cost reductions in silicon production could come if the necessity to grow and slice large cylinders of silicon could be avoided. Development work for the Jet Propulsion Laboratory on a process for growing continuous ribbons of crystalline silicon is going on at Mobil-Tyco (Figure 18) and IBM. Efficiencies as high as 11 percent have been demonstrated with cells made from the ribbon, but the process introduces unwanted inpurities into the silicon and is not yet as rapid as the traditional method. Several other processes for producing sheets of silicon are also being studied, but all of the ribbon and sheet processes are still regarded as uncertain by most observers, and the current optimism within the photovoltaic research community is not based on their prospects.

There is general agreement that photovoltaic systems designed for use with

Figure 18. A continuous ribbon of crystalline silicon being pulled from a crucible of molten material. The process is known as edge-defined growth and has been used to grow ribbons several centimeters in width and more than 10 meters in length. The key to the process is a specially designed die through which the molten silicon flows by capillary action. The ribbons have a thickness of about 250 micrometers, much thinner than conventional silicon wafers cut from a cylinder. Photovoltaic cells which have been made from the ribbon have shown efficiencies as high as 11 percent. Using presently available techniques, however, the rate at which the ribbons form is too slow for the process to be economical. [Mobil-Tyco]

Figure 19. Experimental concentrating photovoltaic system. Plastic Fresnel lenses focus sunlight onto tiny silicon cells. The array shown here can produce 300 watts in peak sunlight. A tracking system keeps the array pointed at the sun. Because the lenses are less expensive than an equivalent area of silicon cells, the cost of the system is greatly reduced. A unique feature of this system is the relatively small scale of lenses and cells; in small cells, heat is more easily conducted away, preventing rises in temperature that would lower their photovoltaic efficiency. [RCA Laboratories]

a solar collector that concentrates sunlight are likely to accelerate the cost-cutting process. Arrays of silicon cells now cost about $10 per watt of generating capacity in full sunlight. Reduction to about $1 per watt—a cost that is expected to make feasible a broad range of specialized applications—is widely anticipated as early as 1980, particularly for concentrating systems. Cells designed for energy densities 100 times that of normal sunlight (100-fold concentration), for example, do not cost appreciably more than those designed for normal intensities, but they generate 100 times as much power. Thus the cost of the concentrating collector rapidly becomes the limiting factor (as explained in Chapter 4). A report by the congressional Office of Technology Assessment finds that "concentrating systems can be developed which provide photovoltaic electricity in the next few years costing no more than $1200 per peak kilowatt."

The surge of enthusiasm for concentrating photovoltaic systems is based on what appear to be very attractive economics—at least for the short run. Indeed, the record low price to date for photovoltaic arrays is a concentrating system being built with federal funds by the Solarex Corporation, Rockville, Maryland, for the Mississippi County (Arkansas) Community College. The 250-kilowatt installation will cost about $2.75 per peak watt for the silicon cells and $2 to $3 more for the concentrators themselves. Experimental concentrating systems are being designed by many researchers; they will operate at a range of concentrations from 10- to well over 1000-fold. At high concentrations most collectors will require active cooling, because the performance of photovoltaic devices degrades as temperatures increase; this prospect has stimulated consideration of photovoltaic total energy systems that would produce low-temperature heat as well as electricity (see Chapter 5). A variety of innovative concentrating collectors for use with silicon cells have been designed, and some of them are being tested at Sandia Laboratories (Figures 19 and 20).

Below $1 per peak watt, however, things get more difficult, and many observers believe that with silicon, at least, only flat-plate arrays will be able to achieve further price reductions. The energy agency's goal is $0.50 per peak watt for flat-plate arrays of silicon cells by 1986—a figure that program officials acknowledge was initially chosen as an arbitrary estimate of what would be necessary before on-site photovoltaic power could be generally competitive. However, there is growing evidence that it is an attainable goal. Within the last year, according to observers familiar with the semiconductor industry, several of the major companies that are analyzing possible production methods for the Department of Energy program have convinced themselves that $0.50 per watt can be achieved, without any breakthroughs, by simply extending and automating present procedures. Paul Rappaport, director of the Solar Energy Research Institute and a recognized photovoltaic authority, says that the cost goal "is doable" if dedicated processing plants large enough to turn out 50 megawatts of

Figure 20. A concentrating photovoltaic array being assembled. The solar collector is of a type known as a compound parabolic concentrator, made of acrylic plastic. Sun-light striking the upper surface is concentrated onto silicon photovoltaic cells cut into thin strips (the dark rectangles shown here). [Argonne National Laboratory]

generating capacity a year are built. The Office of Technology Assessment study, comparing the Department of Energy cost goals to historic learning curves that describe how prices have declined as production volumes increased for other semiconductor devices, characterizes them as "optimistic but not impossible," provided near-term markets for solar cells can be found.

Even if conventional silicon cells do not achieve their goals, they represent only one of many photovoltaic technologies (Figure 21). In fact, silicon faces stiff competition from a growing array of alternative cell concepts—unconventional silicon designs, thin-film cells, and alternative concentrating designs. Many investigators believe that thin films of cadmium sulfide, amorphous silicon, or other photovoltaic materials will ultimately prove to be far cheaper than conventional silicon. And for concentrating systems, where high efficiency is the crucial parameter, designs based on gallium arsenide are proving much more effective. Still other novel approaches may permit very high conversion efficiencies, very low costs, or both.

Concentrating photovoltaic systems place a premium on the efficiency of the solar cell. Commercial silicon cells average about 11 percent efficiency and laboratory versions have reached 15 percent—both relatively low compared to conventional energy technologies. This accounts for a growing interest in gallium arsenide cells, despite the fact that they are now as much as ten times more expensive than silicon. Varian Associates has demonstrated an experimental gallium arsenide cell that achieves 19 percent efficiency operating with sunlight concentrated to 1735 times its normal intensity. At these extreme conditions the cell produces electricity at a density of 0.24 megawatt per square meter of cell area (Figure 22). IBM has made experimental gallium arsenide cells that attain efficiencies of 22 percent under normal sunlight and 24.5 percent with concentrated sunlight. The cells were fabricated with a novel and potentially inexpensive technique known as epitaxial growth, in which a very thin layer of gallium aluminum arsenide forms on the surface of a gallium arsenide crystal. Gallium arsenide cells have the additional advantage that they can tolerate higher temperatures than silicon, up to 200°C, with only moderate losses in efficiency—high enough to produce both heat and power for some applications.

Still higher efficiencies appear to be possible. One provocative idea being pursued by Richard Swanson at Stanford is to convert the solar spectrum to a form in which photovoltaic cells can make better use of it. This approach, known as thermophotovoltaics, employs a complicated geometry in which concentrated sunlight enters an evacuated chamber lined in part with a refractory material. The light strikes a specially constructed silicon cell and part of it is converted to electricity; much of the rest passes through the cell and is reflected to the refractory material, where the energy is absorbed and reradiated at high temperature, in the process lowering its wavelength slightly. The reradiated energy again

Figure 21. Table of photovoltaic cell types and efficiencies. For concentrating systems the efficiency depends on the degree of concentration, so both numbers are given. [Adapted from *Science*, Vol. 199, p.636, February 10, 1978]

Cell type	Probable maximum achievable efficiency	Maximum measured efficiency	Performance of commercial cells
Standard cells and variations			
Silicon—single crystal	20–22%	15%	10–13%
Silicon—polycrystalline		7–14%	
(Silicon + indium-tin oxide)—single crystal	20%	12%	
(Cadmium sulfide + copper indium selenide)—single crystal	24%	12%	
High-efficiency cells for concentrator systems			
Silicon—interdigitated single crystal at a concentration ratio of 100	26–27%	15%	
(Gallium arsenide + gallium aluminum arsenide) at a			
concentration ratio of 200	25–26%	24.5%	
concentration ratio of 1700		19%	
Thermophotovoltaic	30–50%	26%	
Silicon—vertical multijunction	31%	9.6%	
Multicolor	40%	27%	
Thin-film cells			
(Silicon + hydrogen)—amorphous	15%	5.6%	
(Cadmium sulfide + copper sulfide)—chemical vapor deposit process	15%	9.1%	2–3%
(Cadmium sulfide + copper sulfide)—spray process	8–10%	5.6%	
(Cadmium zinc sulfide + copper sulfide)	15%	6.3%	
(Cadmium sulfide + copper indium selenide)—polycrystalline	15%	6.9%	
Gallium arsenide	25–28%	15%	

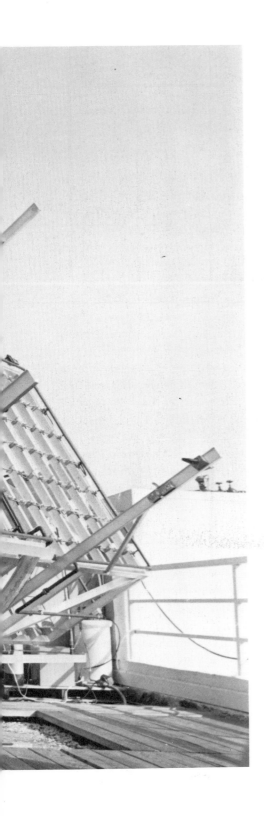

Figure 22. Experimental array of gallium arsenide cells in a concentrating collector. The array tracks to follow the sun. Light strikes the mirrors at the back of the array and is focused into the small cylinders at the front, which contain the still smaller photovoltaic cells (see above). The concentration is 1735 times, producing an extremely intense beam of light. So intense is the beam that the cells operate at elevated temperatures despite a cooling system, but achieve 19 percent efficiency in converting the sunlight to electricity. [Varian Associates]

passes through the solar cell, where an additional portion is converted to electricity on each pass. In effect, the device recycles light until 30 to 50 percent of it is converted to electricity, according to Swanson's calculations with computer models. An early version of an experimental thermophotovoltaic device has achieved 26 percent efficiency.

Less complicated methods of achieving similar results include a range of novel cell designs that amount to fabricating two or more photovoltaic devices in a single cell. Texas Instruments, for example, has developed a high-efficiency design for a silicon cell that would contain two superimposed p-n junctions, and other companies have explored designs in which the two junctions are interleaved or interdigitated. Experimental cells have shown 15 percent efficiency and are expected to reach 20 percent with ease; analytic studies suggest that efficiencies as high as 27 percent should be attainable with concentrated sunlight.

A further step in the same direction is multiple junction cells—two or more separate p-n junctions within a device. Designs in which the junctions are stacked both vertically and horizontally have been studied, and some versions show estimated efficiencies near 30 percent; experimentation with such cells is just beginning. Multijunction cells are more complicated to fabricate but offer a number of potential advantages in addition to their efficiency. Because the junctions are connected to each other in series, for example, the cells produce higher voltages than do ordinary silicon cells, which may simplify interconnection into arrays of practical size.

Still higher efficiencies are considered possible with multicolor photovoltaic cells, in which different photovoltaic materials are used to capture different parts of the solar spectrum, each in its optimum range. In one version studied by investigators at the University of Arizona, optical filters or dyes would be used to partition the incoming light to three separate cells—gallium arsenide, silicon, and germanium. Their analysis suggests that efficiencies of 40 percent or more could be attained with concentrated sunlight. Alternatively, the three different cells might be stacked one on top of another in a single device, with sunlight cascading from one to another. A design developed by Research Triangle Institute, for example, would use only photovoltaic materials from the III–V semiconductor family (which includes gallium arsenide) and would grow each layer of material directly on the layer beneath to create a single unit. The advantage of the III–V materials is that their photovoltaic activity covers a wide range of the solar spectrum and, like that of gallium arsenide, does not degrade very much at temperatures up to 200°C.

An early version of a multicolor cell developed by Varian has shown conversion efficiencies of 27 percent, the highest yet achieved. Since experimental work with multicolor cells and other very high efficiency approaches is in its infancy, the performance and the cost of practical devices are still uncertain. But it is clear that conversion efficiencies above 30 percent, which would put photovoltaics on a par with many conventional energy technologies, would have a major effect on the economics of solar power, particularly for concentrating systems, in which the cost of the cells themselves is not a major factor.

At the other extreme from researchers pursuing expensive, high-efficiency cells are those who believe that flat-plate arrays made from thin films of polycrystalline or amorphous semiconductor materials can be made so cheaply that they are inevitably the way of the future, despite their lower efficiencies. Rappaport, for example, says that "the technology of thin films is still in its infancy" and that thin films may ultimately prove a major competitor not only of concentrating systems but also of conventional crystalline silicon cells. One indicator that buttresses this point of view is the degree of private investment in thin-film techniques and manufacturing facilities.

Optimism about the possibility of dramatic cost reductions with thin-film techniques is based on savings in both materials and manufacturing effort. Materials for a standard silicon cell cost about $150 per square meter, for example, compared to about $3 per square meter for a thin-film cell of cadmium sulfide (Figure 23). Manufacturing costs can also be greatly reduced, since large areas of thin-film cells can be quickly formed by chemical deposition or spray techniques, eliminating the need to grow and slice crystals. These techniques also lend themselves to the incorporation of additional processing steps in the same operation, and much thinner cells, typically a few micrometers or less in thickness, can be formed. The only thin-film cell now close to commercial production is based on cadmium sulfide, for which efficiencies have run 3 to 5 percent, necessitating substantially more cell area than would be required to produce the same power from silicon.

Despite this disadvantage, two firms are now gearing up to produce these cells in large quantities. Solar Energy Systems, a subsidiary of Shell Oil, is preparing to market cells made with a batch, vacuum deposition process at prices competitive with those for silicon cells. Photon Power—originally a subsidiary of the D. H. Baldwin Co. but now primarily owned by a major French oil company—is developing a chemical spray technique in which solar cells are formed directly on hot float-glass. Observers familiar with both processes speculate that cells could be produced in large quantities with either method for a price of about $2 per watt or less—possibly much less if the Photon Power approach can be made to work in a combined facility that would produce both glass and cells.

A major improvement in cadmium sulfide cells was recently announced by scientists at the University of Delaware. They have fabricated experimental cells with efficiencies of over 9 percent. The higher efficiencies were achieved by relatively simple modifications that reduce internal losses of current in the cells

and should be easy to incorporate in a manufacturing process. Estimates by the Delaware group suggest that the more efficient cadmium cells can be produced in quantity for substantially less than $1 per peak watt, although the production processes remain to be proven.

Still other potentially inexpensive thin-film materials are being developed. At RCA, for example, investigators are experimenting with an alloy of amorphous silicon and hydrogen. The hydrogen, which can be added in varying amounts up to about a one-to-one atomic ratio, acts to increase the absorption of light in the film and to improve its photovoltaic properties. The material is produced in a low-temperature process, known as glow discharge, at about 300°C (Figure 24). The amount of silicon required is small (the cell is 1/250 as thick as a conventional silicon cell), and because it is amorphous, it can be deposited on an inexpensive substrate such as glass. The RCA group has made cells with efficiencies of 6 percent and expect to achieve 10 percent efficiencies within a few years. Other investigators are studying thin-film cells analogous to cadmium sulfide cells, but in which indium phosphide or copper indium selenide is also used; these cells have shown laboratory efficiencies of 7 percent.

Another novel approach is a photovoltaic cell developed at Bell Laboratories which combines a liquid and a solid. The design, known as a liquid junction cell,

Figure 23. Thin-film cadmium sulfide cells. The advantage of thin-film cells lies in their ease of manufacture and their potentially low cost. Thin films use much less photovoltaic material than do slices of single crystal cut from large cylinders, and they can be packed more densely than cylindrical cells. Cadmium sulfide cells manufactured by a process similar to the one used for the cells shown here have achieved 9.1 percent efficiency. [Institute of Energy Conversion, University of Delaware]

uses two electrodes immersed in an aqueous solution of certain chemicals (Figure 25). One of the electrodes is a semiconductor material—cadmium selenide in the Bell experiments. When light strikes the semiconductor, current flows from one electrode to the other, much like a wet-cell battery. The chemicals in the solution are not affected by the process. Preliminary versions of the cell have shown 7 percent conversion efficiencies with crystalline electrodes, 5 percent with cheaper polycrystalline forms.

Polycrystalline silicon may also provide another possibility. Solarex, the largest producer of conventional silicon cells, is experimenting with this material, as is the West German firm of Wacher Chemical. The goal is not a true thin-film cell; instead, blocks of polycrystalline material are cast and cell wafers are then cut from the blocks (Figure 26). Polycrystalline material has an ordered, crystalline structure but the size of the individual crystals (or grains) is often too small—a few micrometers—to make an effective photovoltaic cell. Solarex, however, has found a technique for casting blocks in which the grains are a few millimeters across, and they have made experimental cells with efficiencies of 10 percent. The advantage of the approach is that it avoids the time-consuming and expensive production of individual crystals and yet retains the advantages of working with well-established silicon technology. The coupling of polycrystalline

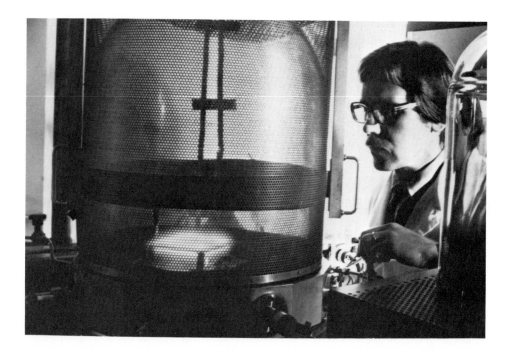

Figure 24. Glow discharge chamber used in making hydrogenated amorphous silicon. A gas composed of a silicon-hydrogen compound is slowly decomposed in the chamber, depositing a uniform thin layer of amorphous silicon containing about 20 percent hydrogen. Because the material is amorphous rather than crystalline, it does not need a crystalline substrate on which to form, and larger cell areas can be formed. Cells from amorphous materials are not as efficient as crystalline cells, but are potentially much cheaper to manufacture; in this case the hydrogen alters the electric properties of the silicon in a way that enhances efficiency. [RCA Laboratories]

Figure 25. Experimental liquid junction photovoltaic cell. The cell replaces the conventional solid semiconductor junction that is at the heart of a photovoltaic device with a liquid-solid junction. Because liquids conform easily to solids, the approach avoids the difficulties of precisely aligning crystal layers as in a conventional cell. In the device shown here, a cadmium selenide electrode forms one part of the junction, a solution of sulfide polysulfides in water the other. The other electrode can be carbon or metal. When light strikes the cadmium selenide, current flows from the cell as if it were a battery. The advantages of a liquid junction cell include inexpensive materials and potential ease of manufacture. [Bell Laboratories]

Figure 26. Experimental photovoltaic cells of polycrystalline silicon. Irregular patterns formed by individual crystals or grains are visible on the cell surfaces. Since the efficiency of polycrystalline cells increases in proportion to the grain size, large grains are preferable. The cells shown here measure 2 centimeters on a side, and cells as large as 5 centimeters have been fabricated. Grain sizes of a few millimeters, such as the largest shown here, are needed to achieve efficiencies of 10 percent. If large-grain cells can be routinely made, they are expected to be considerably less expensive than conventional single-crystal silicon cells. [Solarex Corporation]

materials with new processing techniques—such as ion injection, a method of introducing controlled amounts of impurities into the cells, and laser annealing, which rapidly melts the upper surface of the cell and thus heals crystal damage caused by the ion injection process—is thought by some observers to offer a means of producing low-cost cells in the near future.

The multiplicity of the unfolding photovoltaic options suggests that the federal research program has focused its efforts too narrowly and that its ambitious goals may actually prove conservative. Certainly there are now many possible routes to achieving practical photovoltaic power systems and the limits are far from known. In fact, the sheer richness of the technologies and the rapidity with which new designs and processes are being invented are something of a hindrance to the immediate commercialization of photovoltaic power because there is still no consensus as to which approach will win. Private companies are understandably cautious about investing in production facilities that may prove obsolete.

There already exists a market for terrestrial photovoltaic power systems. In addition to government purchases, silicon cells are being bought for such applications as supplying power to remote Forest Service watchtowers and cathodic protection of pipelines against corrosion. A study done by the Department of Defense for the Federal Energy Administration forecasts a substantial near-term market, eventually as large as 100 megawatts per year, at remote military installations. The study also says that photovoltaic systems for such applications are competitive even at current prices.

The basic unit for photovoltaic power systems is an array of cells producing up to a few tens of kilowatts. Even large central power stations, if they were constructed, would be built up from units of this size. Thus photovoltaic systems are inherently modular, perhaps more so than any other solar technology. Engineering studies conducted for Sandia Laboratories indicate that there is a substantial residential market for photovoltaics, for example, and that there is no technical reason why they cannot compete with other sources of electricity on all scales. Nonetheless, the dominant line of thought within the Department of Energy program has been that photovoltaics can have a major impact only if large, utility-scale applications can be found. "My view," says Henry Marvin, until recently director of the department's solar energy division, "is that the only way to get the cost down is to service some large installations—megawatt size." One knowledgeable critic describes this approach as "a misapprehension; distributed applications have to be the way of the future." Because photovoltaic technology is so modular, however, the department's centralized bias does not yet appear to have affected the technical choices made within the program to the extent evident in other solar programs.

One major on-site application now being actively studied by industry is in electrochemical plants, for which the low-voltage, direct current produced by photovoltaic cells is ideal; because the plant could adjust production to changes in the amount of sunlight available, storage of electricity would not be necessary. Observers familiar with the electrochemical industry say that this market could amount to several thousand megawatts. Given a market, many observers believe that one or more manufacturing plants, each capable of producing as much as 100 megawatts of photovoltaic capacity per year, could be built within 2 years.

There are some signs that the government is beginning to give more attention to photovoltaics. A proposed addendum to the fiscal 1979 budget would increase photovoltaic research by $30 million, and the Congress is considering a bill that would result in spending $2 billion over 5 years to stimulate the photovoltaic market through federal purchases. With or without the government, however, it seems likely that photovoltaic power systems will be an important new entry into the commercial energy sweepstakes well before the end of the century. Knowledgeable observers of the semiconductor industry, such as John Linvill, chairman of the Stanford Electrical Engineering Department, and Lester Hogan, vice-chairman of Fairchild, put it this way: "We believe that photoelectric conversion of solar energy can be made viable as a source of power for terrestrial use within a decade."

4

CAPTURING SUNLIGHT
A Renaissance in Collector Design

Solar collectors are perhaps *the* characteristic solar technology, and much of the national enthusiasm for solar energy has been directed toward developing ingenious designs. The result has been an explosive proliferation of collector types—of metal, glass, plastic, concrete, and combinations of these materials. Although it is only in the last 5 years that a modern technology of capturing sunlight has taken form, literally dozens of designs have been proposed, each offering some particular advantage of manufacture or application. Collector designs range from low-cost, low-temperature versions to devices capable of concentrating sunlight 10,000 times and reaching extremely high temperatures. There is no sign that this upwelling of creativity has abated or that the limits of collector technology have yet been reached.

The economics of capturing sunlight is in flux, too, since an industry to manufacture solar collectors is both young and growing rapidly. A recent government survey counted nearly 200 companies now active in the business, and production has been increasing by 168 percent a year since 1975, doubling roughly every 7 months. Output of flat-plate collectors in the United States amounted to 165,000 square meters during the first 6 months of 1977. The Department of Energy estimates that as many as 24,000 homeowners installed solar energy systems during this period. Many of the newer and more efficient collector designs are just now coming into manufacture and prices do not yet reflect the advantages of mass production.

The ubiquitous flat-plate collector of metal and glass is the type familiar to most people (Figure 27). A few years ago it was the only type of collector available and it is still the commercial leader. But flat-plate collectors are facing growing competition. They are being overtaken in volume by tubular plastic or synthetic rubber collectors designed to heat swimming pools—reflecting in part a ban on the use of natural gas for that purpose in most of the Southwest. Flat-plate collectors are also being challenged in price by passive solar heating systems and by advanced collectors that use cheaper materials and perform more efficiently. Moreover, the advanced collectors are making possible commercial and industrial applications of solar heat at temperatures beyond the reach of flat-plate devices.

The classic flat-plate collector consists of a black metal absorber enclosed in an insulated box with a glass or plastic cover (Figure 28). Collected heat is transferred to air or a liquid and piped to its destination, usually a storage tank containing rocks or a liquid. Variations among more than 100 manufacturers include the type of metal absorber used, the amount of insulation and glazing, and whether the metal surface has been treated to enhance its light-absorbing and heat-retaining properties. Flat-plate collectors work more efficiently in summer than in winter, and more efficiently when producing low-temperature heat than high-temperature heat. Because their performance is so sensitive to climate, efficiencies can range from 70 percent in warm weather to less than 10 percent in very cold weather. Typical applications and temperatures include space heating through "solar assist" for a heat pump (15°C), swimming pool heating (26°C), domestic water heating (54°C), and direct space heating (70°C). On the Fahrenheit scale, the range is 60° to 160°F.

Flat-plate devices collect both diffuse and direct sunlight. Like most solar devices, they are predominantly used as fuel savers, providing 30 to 60 percent of space heating needs and relying on backup energy systems for the remainder. Prices for the collectors alone vary considerably, but are generally in the range $50 to $150 per square meter; a complete solar heating system for an ordinary house can cost $10,000, however, because of high installation costs. Although prices may fall somewhat as the building industry becomes more familiar with solar heating systems, the potential for cost reductions in the collectors themselves is limited by the amount of metal and other expensive materials used in their manufacture.

For very low temperature applications such as heating swimming pools, some 15 U.S. companies manufacture relatively inexpensive solar collectors consisting of mats of synthetic rubber tubing through which water is circulated. These simple collectors typically operate at about 26°C (80°F); they usually boost water

Figure 27. Flat-plate collectors on a house in Quechee, Vermont, supply hot water for domestic use and to assist a heat pump; the pump extracts heat from the hot water instead of the frigid outside air. The house has an area of 2300 square feet. It has 6 inches of insulation in the walls, triple glass windows, and an oil-fired backup heating system. The solar system (including backup) cost $9400 and is expected to supply 40 to 50 percent of the heating requirements, reducing operating costs to half of those for a completely oil-heated house. [Grumman Energy Systems, Inc.]

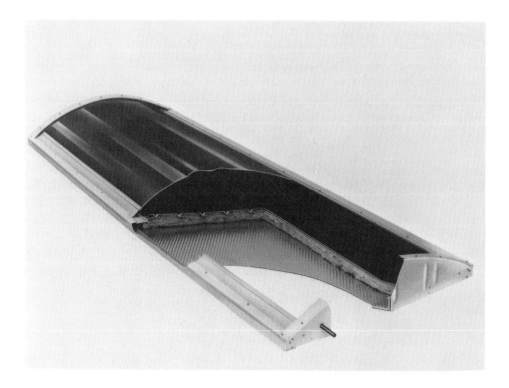

Figure 28.   Cutaway of a flat-plate collector showing the essential components. Sunlight enters the collector through a cover plate made of glass or plastic; it is curved acrylic in the model shown here. A double layer of glass is used for additional insulation in collectors to be used for higher-temperature applications. The absorber is metal, in this case consisting of blackened aluminum strips locked around copper tubes, through which flows the heat transfer fluid. Insulation and a structural frame complete the collector. [Grumman Energy Systems, Inc.]

temperatures by 2° to 5°C (5° to 10°F). Some 280,000 square meters of these collectors were produced in the first half of 1977.

Researchers at Princeton University are developing a related but more sophisticated collector consisting of multiple layers of thin, flexible plastic films—a plastic "flat-plate" collector. Air-filled pockets at the top and bottom would provide insulation and water would flow through a center channel. According to Theodore Taylor of Princeton, composite plastic collectors of this type made from a combination of Tedlar, polyvinyl chloride, and Teflon films have raw material costs as low as a few dollars per square meter and can produce temperatures as high as those attained with ordinary flat-plate collectors. The actual manufacturing costs and the durability of such advanced plastic collectors remain to be determined. Rigid plastic flat-plate collectors are also being developed at Battelle Memorial Institute in Columbus, Ohio, and elsewhere.

For space heating of homes and commercial buildings, flat-plate collectors face competition from passive solar systems. The name arises because heat in passive systems is collected and distributed, without the use of pumps and fans, by means of natural radiative, conductive, and convective processes. This simplicity and the potential for low cost—the building itself typically serves as the solar collector—are the main attractions of passive systems.

The essence of passive solar systems is careful design that works with the surrounding environment to capture and retain heat in the winter and to remain cool in the summer. Thus, windows facing north are frequently made small or eliminated to a large extent, while those facing south are made large but are protected from the summer sun by an overhanging roof. In addition, passive houses frequently include heavy masonry walls or other sources of thermal mass that can absorb the sun's heat during the day and reradiate it to warm the house at night, and thus act as a sort of concrete collector. An early example of these techniques is Montezuma's Castle, built around A.D. 700 by cliff-dwelling Indians in what is now Arizona; the castle sits in a recess of a south-facing cliff, its massive adobe walls warmed in winter but shaded in summer.

At least five distinct techniques are used in modern passive systems, alone or in combination. The simplest is direct gain through extensive south-facing windows with interior walls or floors providing storage. In a well-insulated building this is often enough to supply more than half of the heat requirements. A classic example of direct gain is the Wallasey School in Liverpool, England, a two-story concrete structure with a south wall of double glass windows; electric lights and the body heat of the students provide the only supplementary heat. Direct gain buildings, however, often experience wider variations in temperature than are normally considered comfortable.

One way to limit temperature swings in a passively heated building is to partially decouple the thermal storage from the living space by means of a second

Figure 29. This new 8000-square-foot office building and book warehouse at the Benedictine monastery in Pecos, New Mexico, obtains 95 percent of its heat from a passive
solar heating system, the remainder from electric resistance wall-heaters. The front
(southern) third of the linear building is office space in which worktop counters conceal
two rows of black drums containing 7600 gallons of water. The drums act as a thermal
storage wall. The warehouse portion of the building is heated by direct gain through
clerestory windows atop the building and by a second thermal storage wall of masonry
that separates the two parts of the building. Aluminum reflectors on the sidewalk increase
the heating. The construction cost of the building, including the passive solar system, was
$13.50 per square foot. The monastery calculates that the solar system will pay for itself in
2 years; it is building two more passively heated buildings. [*The Atom,* Los Alamos
Scientific Laboratory]

Figure 30. A solar greenhouse in this home in Santa Fe, New Mexico, collects heat
directly through 400 square feet of thermopane glass. Thick walls inside buffer the living
areas, all of which open into the greenhouse, from temperature extremes. A natural
ventilation system heats an under-floor rock storage bin and also dispells excess heat in
the summer and fall. [Copyright © The New Mexico Solar Energy Association, 1978. All
rights reserved.]

technique, the thermal storage wall. In this design, sunlight entering south-facing windows is absorbed by a wall of masonry (Trombe wall) or water-filled drums (water wall); heat is vented to the building by openings at the top and bottom of the storage wall. The wall protects the interior of the building from high temperatures during the day and transmits its stored energy to warm the interior at night. An office building and warehouse belonging to a Benedictine monastery in northern New Mexico incorporates a water wall passive system that provides 95 percent of the energy needed for heating (Figure 29).

A third type of passive system is the roof pond. Plastic bags filled with liquid are exposed to the sun during the day and covered with an insulated panel at night, when they radiate stored heat to the house below. In summers, this cycle is reversed to provide cooling. A fourth type of system warms by circulating a fluid (air or water) in a natural convective loop. Collectors placed lower than the living space warm the fluid, which rises and carries heat to a storage unit within the house. A fifth type of passive unit is a greenhouse built into or against a house and often separated from the living space by a thermal storage wall (Figure 30). In addition to heat, the greenhouse can supply humidity and food.

Passively heated buildings in the United States number in the hundreds, most of them custom-built. The techniques are not yet common practice in the construction industry, nor have they received as much analytic and engineering attention as have flat-plate collectors. But preliminary studies by J. Douglas Balcomb and his colleagues at Los Alamos Scientific Laboratory, New Mexico, indicate that passive systems can equal or exceed the performance of flat-plate systems of comparable collector area. Balcomb finds that the optimal thickness of concrete storage walls is 30 to 40 centimeters. The passive systems need relatively more thermal storage but can be considerably less expensive overall. Innovations in passive design are continuing. They range from several types of movable insulation for shielding glass areas at night to new and more compact thermal storage systems, such as ceiling tiles developed at the Massachusetts Institute of Technology. The tiles contain a material that undergoes a phase change at 23°C (73°F), storing heat as it melts and later releasing the heat to the room as it solidifies.

If passive systems represent the low-technology challenge to flat-plate collectors, evacuated tubes represent a high-technology competitor. Evacuated tubes can potentially give "twice the performance at half the cost" of flat-plate devices, advocates say. The design consists of an inner glass cylinder blackened to absorb sunlight, enclosed within an outer protective cylinder, with the space between the two cylinders evacuated (Figures 31 and 32). The inner cylinder is usually coated with a material that cuts the energy lost through reradiation; the heat is transferred to a fluid, either air or a liquid, that flows through the inner cylinder.

One interesting variation on the evacuated tube collector is a corrugated

Figure 31. One version of an evacuated tube collector has a clear glass outer cylinder, a dark inner cylinder, and a small central feeder tube to bring heat transfer fluid up into the collector. The fluid then flows along the inside of the inner cylinder, picking up heat absorbed by the collector. The outer face of the inner cylinder is coated so that it is a selective surface, absorbing visible radiation well but emitting thermal or heat radiation very poorly, thus enhancing solar energy absorption. The space between the selective surface and the clear outer cylinder is pumped to a vacuum before the tube is sealed, providing good insulation.
[Owens-Illinois]

Figure 32. Schematic of another evacuated tube collector. The metal fin inside the coated absorber tube aids heat transfer to the fluid. The rule shows rough dimensions in inches. Evacuated tube collectors are manufactured with equipment very similar to that used for making fluorescent lights. [General Electric Company]

Figure 33. Evacuated tube collector on a house near Washington, D.C. According to General Electric, the collector performs significantly better than flat-plate collectors of comparable area and can be mass-produced at a lower cost. [General Electric Company]

glass sandwich developed by Boeing that uses a circulating dark fluid as the absorber and evacuated channels for insulation.

Cylindrical evacuated tube collectors can absorb light coming from any direction (a 360° aperture) and to capitalize on this property they are usually mounted in arrays with a spacing of about one cylinder diameter between tubes and with a reflective material behind them (Figure 33). Like flat-plate collectors, such arrays can make use of both direct and diffuse light, and they work somewhat better than flat-plate collectors early and late in the day. A vacuum provides such effective insulation that evacuated tube collectors are essentially unaffected by either high winds or cold weather, both of which degrade flat-plate performance; in fact, the output of evacuated tubes is essentially independent of the ambient temperature. Their high efficiency, generally between 40 and 50 percent, translates into higher-temperature heat. Evacuated tube collectors usually operate at 80°C or above for space heating and providing industrial process heat, and with reflectors they can operate at 115°C, high enough to drive absorption air-conditioners.

Evacuated tube collectors now sell for about $150 to $200 per square meter, somewhat more than flat-plate collectors. Production volume is still small, however, and the glass collectors appear to have promise for greatly reduced costs. The manufacturing process for evacuated tubes is closely related to that for fluorescent lights and lends itself to automated mass production. In some designs the individual tubes are removable from the arrays, and thus replacing them is like changing light bulbs. The collectors are far lighter than flat-plate devices, use both less glass and less metal (materials are reported to cost about $50 per square meter), and are more resistant to corrosion. Both heating and cooling applications are being vigorously pursued by manufacturers (Figure 34), and evacuated tubes are beginning to be used in some types of concentrating collectors as well.

To reach temperatures much above the boiling point of water with collected solar energy requires more than simply absorbing sunlight, however efficiently. Concentrating collectors increase the intensity of the energy radiating onto the absorber and hence raise the temperatures that can be achieved. Most concentrating collectors use only direct sunlight and must track the sun. Because they are more effective early and late in the day, tracking collectors can capture as much sunlight as flat-plate devices in most parts of the country. However, tracking mechanisms can be expensive and the collectors are more susceptible to wind damage than flat-plate devices. Collectors range from low-concentration designs, which have a concentration ratio of 5 or less and are capable of reaching 150°C, to high-temperature devices with concentration ratios exceeding 100 that can reach 300° to 500°C. In addition to producing higher temperatures and hence more useful energy, concentrators may have economic advantages as well. Heat losses from the absorbers become less important and the mirrors or lenses that

Figure 34. Evacuated tube collectors being used by Anheuser-Busch, Inc., on the roof of their Jacksonville, Florida, brewery to pasteurize beer. The collectors supply heat at 93°C (200°F) to a storage system, which in turn supplies the industrial process heat for pasteurization as needed. The system is the first practical application of solar energy in the U.S. brewing industry. Owens-Illinois, the manufacturer, has sold tens of thousands of square feet of such collectors for use in schools, office buildings, and home heating systems. [Owens-Illinois]

concentrate the sunlight are usually less expensive than a comparable area of flat-plate collectors.

Although dozens of designs for concentrating collectors have been proposed, three generic types can be distinguished, corresponding roughly to low, intermediate, and high degrees of concentration. These are nonfocusing concentrators, trough or line-focusing concentrators that track the sun by rotating along one axis, and two-axis tracking concentrators. The designs now under development include concentrators made from polished metal, metallized glass and plastics, and composite materials. A wide range of absorber materials can be used with these collectors.

Nonfocusing concentrators have the advantage that they do not have to continuously track the sun and do not require the optical precision of a system that must focus the sun's image on the absorber. They can use diffuse as well as direct light and thus can operate on hazy or partly cloudy days, an advantage in midwestern or eastern parts of the country; focusing concentrators, however, get much more sun in the early morning and late afternoon.

The cheapest kind of nonfocusing concentrator consists of a stationary mirror or reflector placed next to a flat-plate collector. Reflectors placed behind evacuated tube collectors also provide a small degree of concentration. The most sophisticated nonfocusing concentrator, known as the compound parabolic concentrator (CPC), was developed at Argonne National Laboratory in Chicago.

The CPC is derived from light-concentrating devices used in high-energy physics. It consists of parabolic surfaces shaped to deliver the maximum amount of light to the absorber for a given concentration ratio (Figure 35). One version that concentrates sunlight 1.8 times can operate as a stationary collector and reach 120°C (250°F). Versions with higher concentration ratios (3 to 6) often use an evacuated tube as the absorber and can operate between 150° and 230°C; at a concentration of 6 the orientation of the collector must be adjusted once a month, but it does not require daily tracking. However, CPC designs require more reflective surface than, for example, a trough collector with the same aperture but a higher concentration ratio; the CPC designs may thus have higher manufacturing costs. Experimental CPC collectors have been made from metal, metallized plastic, and even solid acrylic (Figure 36). None are yet in commercial production, but costs for some designs have been estimated at around $270 per square meter.

More rapid commercialization and lower costs have been achieved with a variety of one-axis tracking concentrators. More than a half-dozen companies are now manufacturing collectors of this type with concentration ratios in the intermediate range (10 to 100) and operating temperatures between 95° and 300°C. Many of these have been developed without government support and preliminary versions sell in the range of $100 to $200 per square meter, including

Figure 35. Schematic diagram showing how solar radiation is concentrated onto the surface of a collector tube by a compound parabolic concentrator (CPC) of cusp design. Because of its shape, the CPC collects light coming from a wide range of angles. Devices of this design need not track the sun. [Argonne National Laboratory]

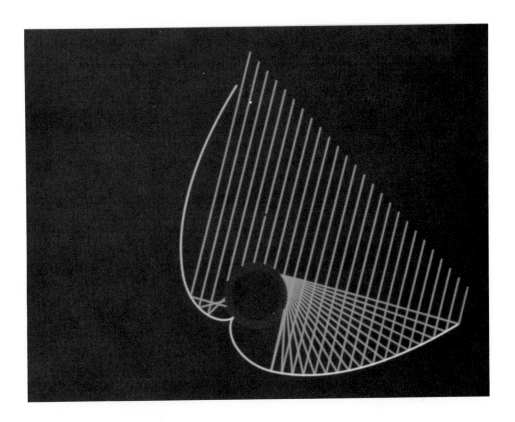

motors and other equipment for tracking the sun; most deliver more than 50 percent of the sun's energy to the heat transfer medium in the absorber. Still other collectors are under development. Installation costs, as with flat-plate collectors, can more than double the final bill, although the wide diversity of costs at present seems to reflect the newness of the industry and the variety of designs employed. Most companies project a cost below $100 per square meter in mass production.

Mirrors arranged to form a parabolic trough that focuses sunlight onto a linear absorber constitute one type of tracking concentrator (Figure 37). The mirrors are typically made of polished metal or coated plastic, the absorbers are made of blackened metal pipe or evacuated glass tubes, and the entire assembly rotates to track the sun from east to west. The applications range from heating and cooling buildings to providing industrial process heat and irrigation pumping for agriculture. The Albuquerque Western company, for example, is marketing a low-temperature version, primarily for home heating, that costs about $100 per square meter, a price that makes it competitive with many flat-plate collectors. A higher-temperature version made by the Acurex Corporation, the Hexcel Corporation, and others can go as high as 300°C (600°F). The Acurex collector is being used to drive Rankine cycle heat engines for pumping irrigation water at

Figure 36.   Compound parabolic concentrators made of solid acrylic designed for use with photovoltaic cells. The concentrator makes use of a phenomenon called total internal reflection to direct light entering its upper surface to the tiny cell at the tip of the cone without significant loss. Thus, the photovoltaic cell generates nearly as much electricity as it would if it were as large as the upper portion of the cone. [Argonne National Laboratory]

several locations in the Southwest, and to supply industrial process hot water at a Campbell Soup factory in Sacramento, California (see Chapter 5).

A second type of steerable trough collector uses plastic Fresnel lenses to focus sunlight on the absorber, achieving the same effect as a parabolic mirror with a smaller optical surface (Figure 38). A version of this concentrator developed with $250,000 of its own money by Northrup, Inc., a small heating and cooling company in Texas, produces 120°C heat with 65 percent efficiency. The collector is being used in heating and air-conditioning applications at a hotel in the Virgin Islands, a university in Texas, and many other locations; the company has orders for more than 10,000 square meters of collectors and is developing an advanced unit to produce higher temperatures.

Still a third type of one-axis tracking concentrator makes use of a fixed trough consisting of flat mirrors arranged in strips and held in a metal frame (Figure 39). Tracking is accomplished by moving the absorber. The firm of Scientific-Atlanta is marketing such a concentrator, designed to reach 300°C, for about $150 per square meter. The design, also known as a Russell collector, was developed by General Atomic, which is testing and selling research versions that can reach 480°C.

The highest degree of concentration and the highest temperatures are

Figure 37. This parabolic trough concentrator is made of aluminum honeycomb with a reflective film of aluminized acrylic. The parabolic mirrors focus direct sunlight onto an absorber pipe of copper or steel that is coated with a selective surface to improve its absorbing and heat-retaining properties. The air or oil pumped through the pipe to collect the heat reaches temperatures of up to 315°C (600°F). The concentrator is 9 feet across and 20 feet long and is controlled by a phototransistor tracking system; at night or in high winds the collector turns upside down for protection. This particular design delivers 55 to 62 percent of the incident solar energy at noon to the heat transfer fluid, according to tests at Sandia Laboratories. It is thus one of the most efficient designs tested. [Hexcel Corporation]

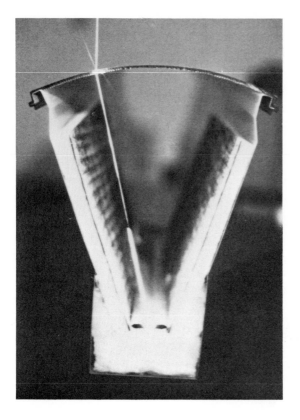

Figure 38. Laser beam refracted by a Fresnel lens as it enters a concentrating trough collector. The lens is made of extruded acrylic that is resistant to damage by ultraviolet light. Sunlight entering the lens is directed to a small copper absorber tube at the bottom of the trough, which is made of galvanized steel insulated with fiber glass. The absorber tube is coated with a black selective surface. Up to 24 collectors of this type can be driven to track the sun by a small electric motor and two tracking photocells. [Northrup, Inc.]

achieved with two-axis tracking collectors and heliostats that reflect light to a central tower (see Chapter 2). Because of the high concentrations, usually 100 to 1000 or more, extreme accuracy in pointing is required and the tracking mechanisms are more elaborate and generally more expensive than those for one-axis systems. However, these systems are very efficient converters of sunlight to heat and can reach temperatures high enough to power conventional electricity-generating equipment.

One of the few two-axis systems now being manufactured commercially is that of the Omnium-G Company (Figure 40). It consists of a tracking parabolic dish similar to those used in radar installations and is operated as a total energy system, producing both electric power and hot water at about 82°C. Energy storage is provided by a compressed-air system for electricity and a hot-water tank for heat. The system sells for about $1000 per square meter of concentrator; the price corresponds to about $2 per watt of installed generating capacity. An alternative approach is to use a fixed mirror and a tracking absorber, as in the system being developed by E-Systems, Inc. (Figure 41). The company is designing versions of the concentrator as large as 90 meters in diameter and projects costs for mass-produced systems as low as $50 per square meter. Still a third approach is that of Sunpower Systems Corporation, which is marketing a collector consist-

Figure 39. A fixed reflector trough and a moving absorber are the distinguishing features of this concentrating solar collector. The faceted reflector is composed of many long, flat mirror segments arranged on a concave cylindrical surface. The mirrors reflect sunlight onto a linear absorber that pivots to follow the sun's movement across the sky. The absorber is an array of evacuated tube collectors. [Scientific-Atlanta]

Figure 40. A complete solar power plant is built around this 6 meter diameter tracking parabolic dish that concentrates sunlight 10,000 times. The system uses solar-heated steam to operate a 7.5-kilowatt generator and also produces 30,000 Btu's of heat per hour in the form of 82°C (180°F) hot water. [Omnium-G]

ing of a series of parabolic trough concentrators arranged on a carousel; the system produces heat at up to 200°C and sells for $130 per square meter.

Concentrated sunlight can also be directly converted to electricity with photovoltaic cells. The concentrating collectors used with these cells are very similar to those used for producing heat; the main difference is that the cells replace the absorber. With concentration factors of 100 or more, the primary cost of such a system is that of the concentrating collector. Hence the development of advanced concentrators will have an impact on both thermal and electric applications of solar energy.

The staggering array of options for collecting sunlight—passive, flat-plate, and concentrating systems—is testimony both to the inventiveness of solar engineers and to the wide range of potential applications for solar energy. Prices now range almost as widely as collector systems. Most analysts expect that prices will decrease somewhat as production increases. More striking is what appears to be a convergence of projected prices in the range $50 to $100 per square meter for flat-plate, evacuated tube, and concentrating collectors—independent of operating temperature over a wide range. If these projections prove correct, higher-temperature applications may become a prime user of heat from the sun. But predictions of any kind are difficult because new collectors are still being designed, new materials are being developed, and the integration of solar collectors into practical energy systems is just beginning. What is certain is that solar collectors are a booming field of technology. If the past few years are a guide to the future, sunlight will be collected in myriad ways.

Figure 41. The experimental concentrator system shown here uses a fixed spherical mirror. Sunlight striking the mirror is focused on a conical absorber that tracks the sun's movements in mirror image. Larger versions now being studied might reach 90 meters across and be set partially into the ground. [E–Systems, Inc.]

5

COLLECTOR SYSTEMS
The Promise of Intermediate Temperature Uses

Among solar technologies that produce thermal energy, most of the attention and funding has gone either to low-temperature systems that are used for home heating or to high-temperature systems that provide centralized generation of electricity. The low-temperature systems are characterized by the utmost simplicity. The high-temperature ones are characterized by a great deal of sophistication and engineering complexity. In between the two, there is a regime of intermediate temperatures that can be provided by solar energy systems in which the optical quality standards are only moderately demanding, yet the efficiencies are markedly superior to those of the simple collectors used for space and water heating. Systems using collectors capable of operating in this regime are much more flexible than low-temperature systems. They can produce heat and electricity in tandem for residential and commercial use, they can provide hot water, hot air, and steam for industrial processes, and they can perform special tasks needed in remote areas, such as pumping water for irrigation. Although test demonstrations of intermediate-temperature systems are only starting, a number of analysts are beginning to conclude that there is so much demand for energy in the intermediate-temperature regime that such systems hold the greatest potential for solar utilization by the year 2000.

One of the largest potential applications is for the production of industrial process heat, an endeavor that consumes 23 percent of the energy used by the United States. Not all of these industries require heat in the high-temperature range that characterizes steel production or copper smelting. The food, textile, and chemical industries all use large amounts of energy in an intermediate-temperature range, which can be roughly defined as 100° to 300°C (212° to 572°F). Thirty percent of all industrial process heat is now used at temperatures below 300°C, and concentrating collectors are presently available that can operate at temperatures up to 300°C (see Chapter 4 and Appendix 2).

Several factors besides temperature matching make solar energy attractive for industrial uses. Large segments of the food and textile industries are already located in regions that are quite favorable for solar energy. Furthermore, the economic case for industrial use is enhanced by the fact that industrial energy demand tends to be relatively constant year-round, and industrial organizations have a professional maintenance staff in-house with the skills necessary to operate a solar energy system. Finally, industry tends to be more responsive to changes in energy economics than the residential and commercial sectors, can act quickly to install solar energy systems if competing fuels become exorbitant or difficult to obtain, and could become a primary user of energy from the sun.

A study prepared for the Department of Energy by ITC Corporation concluded that by the year 2000, intermediate-temperature systems could displace 7.5 quadrillion Btu's (quads) of energy now supplied by fossil fuels for use below 300°C. Industries that require higher temperatures could use solar energy to preheat water, air, or steam used in their processes. Adding together the applications whose terminal temperature falls in the intermediate-temperature range and those that could benefit from solar preheating, ITC concluded that solar energy could supply 36 percent of industrial process heat by 2000.

Concentrating collector systems could also be used for a variety of other applications. In addition to crop irrigation and decentralized generation of electricity, concentrating systems can be used for the combined production of heat and electricity, which the federal energy research establishment has named total energy. Projections of the energy that could be supplied by 2000 from these systems are not available, but the potential market includes much of the irrigation in the West and a substantial fraction of the residential (and commercial) heat and electric supply, especially in the "sun belt." (Concentrating collectors that operate at intermediate temperatures are also well-suited to run absorption air-conditioners.) Another feature of concentrating collectors is that their scale is naturally suited to small and medium-sized installations, and the hardware is varied enough that it can be easily tailored to different applications. Because small systems can be built and installed faster, intermediate-temperature systems have the potential to supply substantial amounts of solar energy much more rapidly than systems on the scale of (for instance) the power tower.

All the pieces needed for intermediate-temperature systems exist now. At least ten varieties of one-axis tracking collectors are now being made in the United States, at a cost—before installation—of $100 to $200 per square meter, and most companies project a cost of about $100 per square meter in mass production. The collectors, which are generally parabolic troughs that focus sunlight onto a dark absorber, are already approaching the costs of simpler flat-plate collectors used for space and water heating at temperatures below 100°C.

For space heating, for industrial processes, and for operating absorption air-conditioners, concentrating collector systems operate by circulating the heat transfer fluid (generally water or oil) through an absorber and then through the other components in the system. To produce electricity, the fluid is often circulated through a small heat engine which looks somewhat like a gasoline engine but which derives its operating energy from the solar-heated fluid rather than from burning gasoline. The development of heat engines is less advanced than that of solar collectors, but examples of working devices already on the commercial market are available. European firms are frequently ahead in this field, and so much American engineering effort has been devoted to large-scale steam turbines in recent decades that the status of domestic small heat engine development has been called archaic. But a number of American firms are producing and selling prototypes, and one or two companies are preparing to set up production lines.

Another approach to the production of electricity (and heat) is to use photovoltaic cells in conjunction with concentrating collectors to directly convert sunlight to electricity. The conversion efficiency of photovoltaic cells is relatively insensitive to the concentration of sunlight upon them, and the cost per unit of power drops dramatically when a relatively cheap collector is used to focus sunlight onto a relatively expensive photovoltaic cell. Cells made of silicon are suitable for concentrating systems, and cells made of advanced materials such as gallium arsenide have additional advantages because they can operate at higher temperatures and efficiencies. In either case, the photovoltaic cell replaces the absorber, and the collected sunlight that is not converted to electricity becomes heat—a simple form of total energy system. Concentrating photovoltaic systems can be based on either one-axis or two-axis tracking collectors, but because the cells operate best when they are evenly illuminated, it appears that the two-axis type is preferable (see Figures 6 and 22). The unconverted energy from concentrating photovoltaic systems is generally at the lower end of the intermediate temperature range, 100–200°C.

Although the pieces needed for systems are already available, very few institutions have put them together. In the United States there are two irrigation systems, one total energy test-bed, less than a half-dozen industrial process heat projects, and a larger number of solar air-conditioning systems that use tracking

Figure 42. The Campbell Soup Company plant in Sacramento, California, uses flat-plate and concentrating collectors to produce hot water for its canning operations. Three rows of parabolic concentrating collectors boost the temperature of water from the flat-plate collectors to 88°C (195°F). The concentrating collectors, built by Acurex, are coupled in groups of eight, driven by a single drive system to track the sun. Each unit is 10 feet long. Hot water is stored in a 20,000-gallon insulated tank. The system was designed several years ago when concentrating collector prices were higher, and would probably use concentrating collectors throughout if it were redesigned today. [Acurex Corporation]

collectors. The power tower so thoroughly dominates the federal solar research program (and particularly the subprogram for solar thermal research) that relatively little support has been provided for intermediate-temperature systems—in fiscal 1979, they received only $28 million while the power tower received $70 million.

Furthermore, the research effort for intermediate-temperature applications is fragmented. Work on improved heat engines is done by the conservation directorate. Funding for irrigation and total energy systems comes from the solar thermal subprogram. Support for industrial process heat comes from the solar heating and cooling program, and work on photovoltaic total energy systems is split between the photovoltaic subprogram and the solar thermal subprogram.

All the intermediate-temperature systems use collectors that concentrate sunlight by a factor of 5 to 60 and therefore draw on the same pool of solar technology. But the federal solar program, particularly during its period of rapid growth under ERDA, has been organized by electricity production classifications rather than solar capabilities, so the various midtemperature applications have been separated from one another in a way that makes the useful exchange of research difficult and gives the cumulative effort very little visibility, either within the energy agency or in the view of the public.

One of the most puzzling aspects of the federal program is the lack of emphasis on solar industrial applications, in spite of the agency's own projections. Industrial projects have been lumped together with agricultural projects to make a small demonstration program that has little or no mandate for research. The managers of that subprogram have been terribly parsimonious toward industrial projects, putting most of their money instead into agricultural projects with flat-plate or evacuated tube collectors. This preference is particularly hard to understand when the agency's own projections (circa 1976) are that solar agricultural applications could displace 1 percent of the country's fuel usage by 2000 and solar industrial applications could displace 4 percent. Of 72 projects undertaken in 1977, the subprogram had only 3 that used concentrating trough collectors. One of these is the Campbell Soup plant in Sacramento, California, where parabolic concentrating collectors, used together with flat-plate collectors, produce hot water at 88°C (190°F) for washing cans (Figure 42).

An application with much greater potential is the production of industrial steam. According to a report by the Battelle Columbus Laboratories, process steam is the dominant form in which thermal energy is used by American industry. The Battelle report set a very conservative temperature limit (350°F) in surveying industries that could use solar energy, but even so found that 2.6 quads of industrial steam could be produced by solar each year. But one-axis tracking concentrators can reach much higher temperatures (see Appendix 2).

An example of the way concentrating collectors can be used to produce industrial process steam is the system being designed for a Johnson and Johnson plant in Sherman, Texas. The Johnson and Johnson system, even though it is one of the first designed for use in the United States, will produce steam at 173°C (345°F). The application is particularly simple. Pressurized water (at 120 pounds per square inch) is heated to temperatures as high as 216°C in the collectors, then pumped into a boiler and "flashed" to produce steam on demand. The boiler provides a limited amount of storage, and the existing steam supply system is retained as a backup for the solar unit (Figure 43). Because energy is produced in the form and temperature range in which it is needed, solar industrial systems have perhaps the highest efficiencies of any solar energy systems. Over 50 percent of the energy in the incident sunlight can be delivered to the desired end use.

Solar industrial energy systems offer substantial benefits as fuel savers, because over 90 percent of the energy they would displace is now provided by oil and gas. So little study of industrial systems has been done to date that their economics are quite ill-defined. Estimates of the cost range from $2 to $13 per million Btu's, while the cost of heat from an oil-fired backup system is about $2.50 per million Btu's (approximately $0.30 for the system and $2.20 for the oil). Because backup systems will generally be retained in any case to provide the high degree of reliability needed for industrial operations, storage is not expected to play much of a role. The economics of solar industrial process heat systems will thus depend mainly on the cost of the collectors, which presently comprises 50 to 60 percent of the total bill, and the special engineering presently needed to design the systems. A crucial factor that only experience can determine is the expected lifetime of the system. (The $13 cost figure appears to be based on a very short 6-year lifetime, while $2 to $5 estimates are based on a 20-year lifetime.) But as the cost of collectors decreases with larger production runs and the cost of system design falls as custom-tailoring becomes less necessary for every installation, the price of solar industrial heat is almost certain to drop.

The application of collector technology to the development of total energy systems is another area of solar thermal research with great potential, but it is one in which the federal program has made little contribution so far. As with the industrial program, the solar total energy program is hampered by a lack of analytical breadth and an underestimate of the importance of the application. The solar thermal program managers have hurriedly put together two ambitious demonstrations, but there is little evidence that they are part of a coherent plan. An analysis of the optimium size of total energy systems in a range of applications in different parts of the country is particularly needed. Nevertheless, the total energy effort reflects the energy agency's preference for ever larger systems.

Agency spokesmen doubt that the optimum system would produce less than

Figure 43. Schematic of a solar industrial system designed to provide process steam. Generating steam with a concentrating collector is one of the most straightforward applications of solar technology. There is only one fluid loop in the solar part of the system, and pressurized water is stored in the boiler tank, which also provides steam on demand by controlling the vapor pressure through a steam control valve. No heat exchanger is needed in the system. [Acurex Corporation]

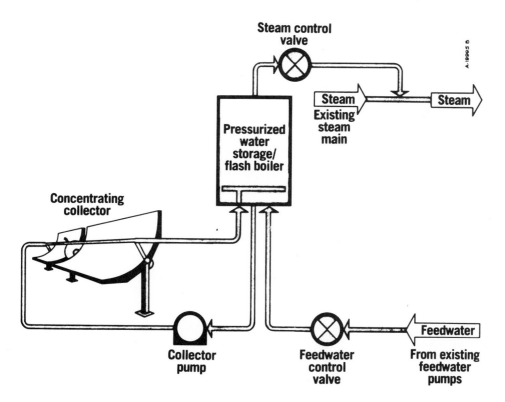

200 kilowatts of electricity (about enough for 20 homes), and the program includes planning for a 1-megawatt electric system in 1981. It also includes a slot for a very large total energy system, which would produce 10 megawatts of electricity and 50 megawatts of heat, scheduled for 1985. The proper balance between electricity and heat in a total energy system is also being studied. The field responsibility for testing total energy systems has been delegated to Sandia Laboratories, where researchers are currently operating a 32-kilowatt system that generates electricity with an organic Rankine cycle heat engine and produces heat for use in the laboratory office buildings. More of a test-bed than a functioning system, the Sandia facility is used to compare the performances of various heat engines and concentrating collectors, as well as thermal storage systems.

As the planning moves toward large total energy systems, the agency has already approved two $15 million projects that will be built by the end of the decade at Shenandoah, Georgia, and Fort Hood, Texas. Each of the projects will produce 200 kilowatts of electricity and 1.5 megawatts of thermal power. In addition, the Department of Energy is funding a $6 million photovoltaic total energy system for installation at a community college in Blytheville, Arkansas.

The Shenandoah project is particularly interesting because it will produce electricity, hot water, heating, cooling, and process steam for a textile factory employing 150 people. When the factory is completed in 1981, it will be leased to the West German knitwear firm of Wilhelm Bleyle, K.G. In direct sunlight the solar system will produce 1000 pounds of steam per hour at a temperature of 169°C—it is sized to supply all the heat the factory needs. The solar energy system will have 6,000 to 10,000 square meters of collectors when it begins operating in 1981, and there are plans to later double the size of both the solar energy system and the factory. The needs of other textile plants are not unlike those of Bleyle, and most of the textile industry in the United States is located in the Southeast, a choice region for solar development.

The southwestern and western regions of the United States have less of an industrial base, but agriculture is economically important in most western states and irrigation is a necessity. Water pumping with concentrating solar systems is attractive both because irrigation is usually needed at sunny, remote sites, and because the need for water over the course of the year is roughly in phase with the incidence of sunlight. Solar-powered water pumping was one of the first intermediate-temperature applications to get under way, and because of the slow pace of the development of industrial and commercial systems, it is currently one of the best examples of the power of solar technology in the intermediate-temperature regime.

The first and still the largest solar pumping facility in the United States was built not by the government but by a private R&D laboratory with the backing of a large life insurance company. Northwestern Mutual Life had a farm near Phoenix, Arizona, that needed water and Battelle Memorial Institute wanted to build on its experience in solar energy research, so the two organizations undertook a 38-kilowatt (50-horsepower) irrigation project in August 1975.

Within 18 months, the joint project began pumping water at Gila Bend, Arizona (Figure 44). The system has 550 square meters of collector surface, and at the peak of solar insolation in June it can pump 10.6 million gallons of water per day. The collectors are parabolic troughs made of aluminized Mylar by the Hexcel Corporation and the heat engine is a Rankine cycle turbine developed by Battelle, using Freon as a working fluid (Figure 45). The system efficiency is 7 to 9 percent, and the facility operates with very little maintenance except periodic washing of the collectors.

As the first of its kind, the system cost about $250,000, but Frank Dawson at Battelle estimates that in limited production the cost would be $75,000. Battelle and Northwestern Mutual surveyed 17 western states and found that there are more than 300,000 irrigation pumps in use there, operating at an energy cost of over $700 million per year, and most of them have about the same power as the Gila Bend facility (50 to 100 horsepower). Dawson estimates that the world market for such pumps is ten times the U.S. market.

The operating temperature of the Battelle-Northwestern Mutual system is 150°C, considerably higher than that attainable with flat-plate collectors. The thermodynamic advantage that intermediate temperature affords can be seen by comparison with a flat-plate solar irrigation system being sold by the French industrial consortium SOFRETES. The overall efficiency of the SOFRETES system is only 1 percent rather than 7 percent, so it is very expensive. Although the system is reportedly subsidized, a 1-kilowatt version costs about $15,000.

Battelle-Northwestern Mutual is not the only American enterprise that thinks it can undercut the French price. An engineer who has been working on heat engine development since 1968 is selling 10-kilowatt solar irrigation systems for a package price of $40,000. Wallace Minto, who heads Kinetics, Inc., in Sarasota, Florida, has sold three of these systems for use in Sri Lanka, Senegal, and Mexico.

A year or so after the Battelle-Northwestern Mutual project started up, the energy agency started a similar project near Willard, New Mexico. Sponsored by the state of New Mexico as well as by the Department of Energy, the Willard system produces 18 kilowatts (25 horsepower) of pumping power from a field of Acurex collectors (Figure 46). The solar pump is used to fill a holding pond, from which water is pumped out into the fields with a conventional diesel pump. For reasons that are not entirely clear, the system has an insulated tank for thermal

Figure 44. A 10,000 gallon per minute solar irrigation system at Gila Bend, Arizona. The system uses parabolic tracking collectors to collect radiant energy and drive a 50-horsepower turbine. It was privately developed by Northwestern Mutual Life Insurance Company and Battelle Memorial Institute, operated by them for the 1977 irrigation season, and then incorporated into the Department of Energy research program. [Northwestern Mutual Life Insurance Company]

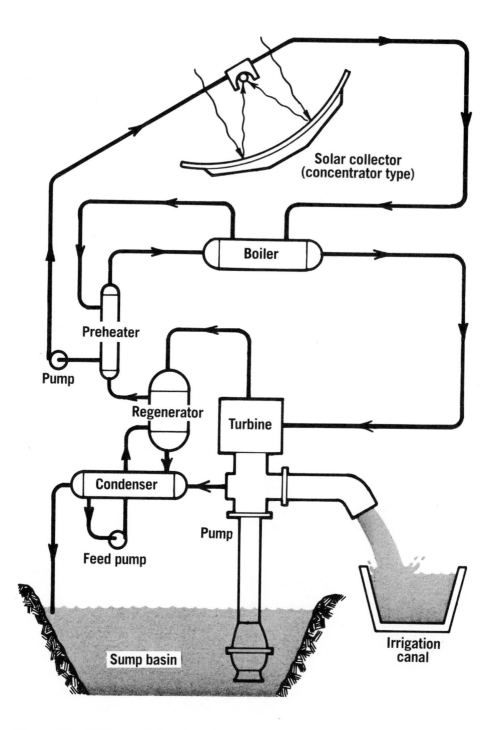

Figure 45. Schematic of the solar irrigation system at Gila Bend, Arizona. Sunlight from the collector heats hot water, which in turn heats Freon in two stages (preheater and boiler) to produce the vapor that turns the turbine. The turbine is coupled directly to a 10,000 gallon per minute pump. Exhaust vapor from the turbine is condensed and recirculated through the system. [Northwestern Mutual Life Insurance Company]

storage, which can be used to operate the solar pump after the sun is down. For an irrigation system, especially one designed with a holding pond, it could be cheaper to store water than heat.

The Department of Energy has plans to build a considerably larger system, also using Acurex one-axis tracking collectors, for pumping water from deep wells (Figure 47). The deep-well system will produce electricity with the heat engine—at some loss of efficiency but increased system flexibility—so that the pump could be located some distance from the collector field. The 150-kilowatt deep-well system, for wells up to 400 feet in depth, will have eight to ten times more collector area than the present systems and will be built in Coolidge, Arizona.

While collector costs can be expected to be reduced by as much as one-half over the next decade or two, the potential savings for heat engines are even greater. The engines used in most of the intermediate-temperature systems to date have been ordered one at a time—and each has usually been handcrafted by a specialized research and development company. The 32-kilowatt engine at Sandia Laboratories (Figure 48) was built by Sundstrand under a handsome government contract. Battelle reports that the engine for its irrigation project cost $50,000, but if it were mass-produced the same engine should cost no more than $3000. Wallace Minto is reportedly setting up a facility to produce a 10-kilowatt Freon heat engine (Figure 49) at the rate of 100 per month, and his company is selling the engine complete with an alternator for $1250 per kilowatt. A number of other U.S. corporations are rushing forward with plans to market small heat engines, but are hesitant to publicize the effort yet. The Office of Technology Assessment notes that if any particular small heat engine were produced at the rate of 10,000 per year, the cost should fall to $200 to $300 per kilowatt. The Jet Propulsion Laboratory estimates that units produced in even larger numbers should sell for $13 per kilowatt, which is about the present cost of automobile engines.

For the near term, one-axis tracking collectors and heat engines of the Rankine type are furthest along and are most suitable for assembling systems. But small two-axis tracking collectors have even greater promise for the longer term because they can produce still higher temperatures. The system now being produced by Omnium-G (see Chapter 4) is the first example to be sold commercially and is still quite expensive. But small dish-type systems are not substantially different from satellite-tracking antennas, and work has only begun on ways to try to reduce costs, particularly through designs amenable to mass production. Whereas the heliostats used for the power tower must be field-assembled, a number of observers of solar technology development believe that there are several attractive ways to achieve factory production of dish collectors. The low-cost plastic collector being built by Boeing (Chapter 2) is one example of this sort

Figure 46. Solar irrigation system assembled by the Department of Energy at Willard, New Mexico, east of Albuquerque. The system uses 625 square meters of collectors and pumps water from a 23 meter (75 foot) deep well into a holding pond. Water is distributed from the holding pond into the fields by a diesel pump. [Acurex Corporation]

Figure 47. Artist's conception of the deep-well irrigation system planned for installation at Coolidge, Arizona. The system will use 4550 square meters of parabolic concentrating collectors and have an operating power of 150 kilowatts. A thermal storage tank with a capacity of 114 cubic meters is included in the system for pumping during cloudy periods and evenings. [Acurex Corporation]

Figure 48. A 32-kilowatt heat engine used in conjunction with concentrating collectors at the Sandia Laboratories total energy system test facility. The working fluid for the engine is toluene, heated to 300°C and entering the engine with a pressure of 275 pounds per square inch. The organic Rankine cycle engine was developed by Sundstrand. [Sandia Laboratories]

Figure 50 (facing page). Comparison of the economics of central receiver solar electric plants with systems producing the same total amount of power from small parabolic dish collectors, each dish being approximately 10 meters in diameter. The study, which was gauged to reflect the expected performance in the year 2000, assumed a cost of $194 per square meter for the small dishes and $170 per square meter for central receiver heliostats. The cost of electricity from small-dish systems proved relatively constant over four orders of magnitude. Not until the size of the central receiver becomes prohibitively large (about 1000 megawatts) does the cost of electricity compete. [Richard Caputo, Jet Propulsion Laboratory]

Figure 49 (above). A 10-kilowatt organic Rankine cycle heat engine produced by a small firm in Florida. The working fluid for the engine is Freon, and the inlet temperature may be as low as 85°C (185°F). [Wallace L. Minto, Kinetics, Inc.]

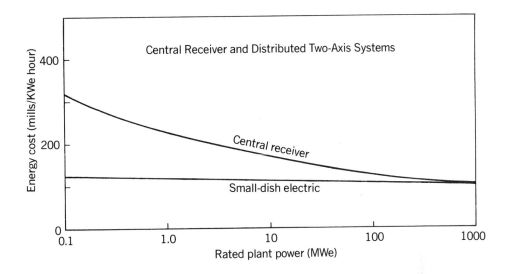

of approach. Another being studied by the Jet Propulsion Laboratory is the use of foam-glass as a construction material.

Small dish-type collectors with concentration ratios of 1000 or higher can produce temperatures up to 800°C. Because of this, they have higher efficiencies for direct thermal applications than one-axis tracking concentrators—70 percent is attainable. The high temperature also makes dish collector systems suitable for heat engines that are capable of higher efficiencies than Rankine cycle engines. The Stirling cycle appears to be very well suited to the 600° to 800°C temperature range.

Small dish collectors with high-temperature heat engines mounted at the focus of the collector may eventually prove to be the cheapest means of producing solar thermal electricity. In a study of distributed-collector methods of producing solar thermal power, Richard Caputo of the Jet Propulsion Laboratory found that systems using dish-type collectors produced electricity that was approximately 40 percent cheaper than electricity from other types of distributed collectors. Perhaps more importantly, the study showed that electricity from small dishes was much cheaper than electricity from power towers in the capacity range (10 to 200 megawatts) being contemplated by the federal solar program (Figure 50). The system study was intended for a plant that would start operation in 2000. Even though the collector cost was estimated to be slightly higher for the small dish ($182 per kilowatt) than for the power tower ($145 per kilowatt), the energy conversion components of the distributed system were superior in efficiency and lower in cost. The Jet Propulsion Laboratory analysis was that electricity from a small-dish electric plant would cost 11 cents per kilowatt hour and the cost for a power tower would be 17 cents per kilowatt hour (assuming the size of each system was 10 megawatts).

Small-dish collectors could be used in total energy systems just as concentrating trough collectors could, and Boeing, among others, is investigating ways to use dish collectors for photovoltaic total energy systems.

Because of the wide range of technology that is already available, intermediate-temperature systems offer great flexibility to perform many functions, and the even greater capabilities of small-scale high-temperature systems may not be far behind. Because the basic units of these systems are small in scale, they can be deployed rapidly. Commercial activity in the development of components is intense, but the demonstration program has been overly cautious and ineffective. Many financial impediments remain for communities or companies that would experiment with intermediate-temperature systems. Some of them could be lessened by tax structure changes, particularly for industries, which can write off the costs of oil but not of solar energy systems.

What is badly needed from the energy research agency is a unified program, free from the shadow of the power tower, that has a mandate to thoroughly analyze the important questions about industrial and other systems that have been neglected so far, and then to build on the advances in collector technology. In the early 1970s, the surprising truism was that residential solar heating systems were ready for use. The largely unappreciated truism of the late 1970s is that the key components for commercial, community, and industrial systems are likewise ready for wider use.

PHOTOSYNTHESIS FOR FUELS
Biomass Energy Rediscovered

Firewood is still a familiar fuel in much of the United States, but somehow it does not come up in discussions of national energy policy very often. Yet wood was the primary U.S. fuel only a century ago and is still the main source of energy in most of the developing world. As recently as World War II, Sweden, cut off from oil imports, derived virtually all its fuel from wood. Since the Arab oil embargo there has been renewed attention in many countries to the energy potential of diverse forms of biomass—wood, sugarcane, algae, and even material produced by artificial photosynthetic processes.

Wood and dry crop wastes have an energy content of 14 to 18 million Btu's per ton, comparable to that of western coals. Raw biomass contains virtually no sulfur and little ash, however, and except for some difficulties in handling it is as easily burned or gasified as coal. Other chemical and biological conversion techniques exist too, most notably fermentation of sugar and grain crops to ethyl alcohol (ethanol), and anaerobic digestion of wet biomass wastes to methane. Thus biomass is potentially a renewable source of a full range of storable liquid and gaseous fuels for which domestic sources of their fossil counterparts are increasingly in short supply.

Biomass-derived fuels may, in fact, have a key strategic role to play in U.S. and world energy affairs. So much of the world's industrial plant and transportation system depends on liquid and gaseous fuels that it is difficult to conceive of a world without oil or natural gas. Yet it is just these fuels that are increasingly in short supply from domestic sources. Coal, the major substitute source of fuels on a worldwide basis, is not evenly distributed geographically—80 percent of recoverable reserves are located in the United States, the Soviet Union, and the People's Republic of China. In contrast, biomass is a more widely distributed source of fuels and one for which many countries have in abundance the essential resources—land, water, and sunshine. Even in coal-rich countries such as the United States, however, biomass may play an important role. The difficulties and costs of obtaining clean synthetic fuels (synfuels) from coal have yet to be confronted in practice, and the sheer magnitude of the effort required to replace oil is staggering. To replace 8 million barrels of oil a day, the amount the United States imported in 1977, will require some 160 large synthetic fuel plants of a complexity rivaling nuclear power stations and costing between $1 billion and $2 billion each. Because the primary alternative to fossil synfuels is biomass fuels, many analysts argue that the proper comparison of the potential of biomass conversion is not to oil and gas at today's prices but to synthetic fuels at probably twice today's prices. In a fuel-short world, biomass may come to be valued as a critical if limited source of liquid and gaseous fuels.

Biomass is already the largest source of solar energy in use in the United States. In recent years nearly a million modern wood-burning stoves and heaters have been sold—an installed energy capacity that outstrips all other direct and indirect solar energy devices (Figure 51). The wood products industry now derives an even larger amount, 40 percent of its total energy needs, or about 1 quad (10^{15} Btu's) from burning bark and mill wastes. Several studies done for the Department of Energy and its predecessor agency, ERDA, suggest that annual production of biomass fuels could conceivably reach 10 quads by the year 2000.

Except for analytic studies, however, the federal energy research program has largely downplayed the biomass option. The ERDA biomass effort initially focused on municipal wastes, a choice that is now generally conceded to have distracted attention from more significant possibilities. The remaining effort has concentrated on the long-range prospects for conversion of biomass to liquid and gaseous fuels and has been meagerly funded—a proposed $27 million in fiscal year 1979 compared to more than $300 million for coal-based synthetic fuels. Direct combustion of biomass and other near-term applications appear to have been neglected.

State and private efforts have been more aggressive. In Vermont, the Burlington Electric Company is building a $40 million new electric power plant that will burn 470,000 tons of wood chips a year. The wood will cost about $12 a ton delivered to the plant and is cheaper than any other fuel available to the utility. In Michigan, Dow Corning is studying a wood-fired boiler that would burn 400,000

Figure 51. Wood-burning furnaces have been used in one form or another to heat houses since the invention of the Franklin stove. Space heaters and stoves designed for wood have been experiencing a strong comeback since the oil embargo, especially in New England and the upper Midwest, where winter fuel bills make cutting your own fuel seem especially attractive. Sales are now close to a half-million units a year. Modern designs incorporate a mechanism for preheating air entering the furnace and then permit the recirculation of combustion gases to achieve more complete burning. The unit shown here is thermostatically controlled with a bimetal coil that opens and closes a damper. [Riteway Manufacturing Company]

tons per year to provide both power and steam to their manufacturing plant. In Saugus, Massachusetts, a plant that burns trash and other municipal wastes from nearby towns and generates steam for making electricity is already operating (Figure 52).

In California, the California Energy Commission has revived a gas producer copied directly from old Swedish designs. According to Robert Hodam of the energy commission, the gasifier is capable of converting nearly any kind of dry agricultural or wood wastes to low-Btu gas with an efficiency of 80 to 85 percent. The gas can be burned in a boiler in place of oil or natural gas. Although ERDA rejected a request to co-fund a demonstration of the device on the grounds that it was not sufficiently novel, the energy commission went ahead. A test unit at a Diamond/Sunsweet Inc. walnut factory near Stockton, California, has operated so successfully on walnut shells that the company has decided to build a larger (130 million Btu per hour) gasifier to provide all its energy needs (Figure 53). Based on bids already received, the company expects the gas to cost about $1 per million Btu's, less than half the cost of the natural gas it now burns.

The demonstration has generated considerable interest in dozens of other companies with substantial biomass wastes, and a major farm equipment manufacturer is planning to build and market gasifiers for farm applications. Hodam estimates that in California alone, where already collected lumber and mill wastes exceed 5.5 million tons per year, the gasifier has an immediate potential for displacing 0.1 quad of oil and natural gas.

The overriding question about biomass energy is the extent of the resources that are or could be available. Those that can be used immediately are small compared to the 75 quads of energy now consumed by the United States every year, and biomass cultivation for energy on a large scale may not be economic, at least in the immediate future. Moreover, most agricultural scientists believe that energy uses of biomass must coexist with needs for food and fiber, so that very large areas of prime land may not be available exclusively for energy crops in heavily cultivated countries such as the United States, although they may be in other countries such as Brazil.

Nonetheless, there is increasing optimism among biomass analysts that substantial amounts of energy could be available even in this country from wastes, and from field, tree, and aquatic crops. For one thing, food yields for most crops are only about half the yield of total biomass under a wide range of conditions. Both intensive agriculture and forestry produce substantial quantities of stalks, straw, bark, and other residues. According to a study by SRI International, potentially collectable crop and feedlot wastes in the United States amount to more than 300 million tons per year (dry weight). This corresponds to about 4 quads of energy per year, or almost twice the U.S. on-farm expenditure of energy in fuels and fertilizer. Thus U.S. farmers could in theory decouple themselves from conventional energy sources and become largely self-sufficient—this is a possibility that is attracting new attention. The Swedish forest products industry already produces 60 percent of its energy needs from waste biomass and the larger U.S. products industry may be even closer to energy selfsufficiency by 1990. Thus U.S. wastes and residues alone might in principle provide the energy equivalent of more than 3 million barrels of oil per day by the end of the century. The practical difficulties of collecting these materials and converting them to fuels are substantial, but they are not a negligible source of energy.

A second consideration is that many potential sources of biomass for fuels can be grown on marginal land that is unsuitable for food production or on bodies of water. Trees, for example, will grow on swampy land unsuitable for crops. Several species of plants that produce hydrocarbons grow well in arid regions. Water hyacinth and algae have both been proposed as suitable for freshwater farming, and marine kelp farms are being studied as a way of using the oceans themselves.

A third factor is that many individual countries are so well endowed with land, water, and sunshine that fuels from biomass can play a major role in their internal economies and may even become a source of export earnings. This is particularly true for many countries in the world's tropical and semitropical belts. Brazil, for example, is planning to largely replace imported oil with ethanol made from biomass over the next two decades, and the country also has the physical resources to produce additional ethanol for export should the world energy market warrant it.

Yet it must be acknowledged that when it comes to growing biomass as an energy crop, there are still substantial uncertainties. Energy is not as valuable a crop as food or fiber and cannot compete for prime land. Raw biomass is usually the dominant cost item in most analyses of energy plantations. For the present, at least, the use of agricultural or other residues or the coproduction of energy and food or fiber in an integrated process seems to have advantages—a coordinate rather than a competitive approach.

The key limitation to growing biomass resources on a large scale is the low efficiency of green plants as solar energy converters. Although the photosynthetic process has a theoretical efficiency of 6.6 percent, the highest yields obtained with the most efficient plant known, 112 tons per hectare with sugarcane, correspond to an efficiency of about 3.3 percent. Most plants, even under intensive cultivation produce biomass with an efficiency of less than 1 percent. Clearly there is room for improvement, and it would be shortsighted to assess biomass resources without acknowledging that potential. This is an era characterized by a revolution in understanding the fundamental biology of the cell. Rapid advances in research have begun to lay the groundwork for mastering and perhaps altering the molecular mechanisms of genetic and regulatory systems. Applications of these

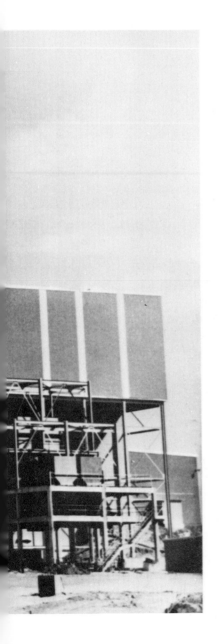

Figure 52. Trash-burning plant near Saugus, Massachusetts, built with private funds to incinerate municipal wastes from ten neighboring towns. The heat is used to produce steam, which is sold to generate electricity for a manufacturing plant. Energy recovery from municipal wastes, common in Europe, is still relatively rare in the United States. [Department of Energy]

Figure 53. Pilot plant gasifier at the Diamond/Sunsweet plant near Stockton, California. The unit gasifies walnut shells that the company formerly paid to have hauled away. The resulting gas is passed directly into a boiler, where it is burned to raise steam and generate electricity. The design of the gasifier, known as a producer, was taken directly from Swedish designs that date back more than a century. The producer gasifies the walnut shells through a process known as partial combustion, in which a portion of the material is burned, releasing heat that volatilizes the remainder. In coal gasification the hot gases are usually cooled before combustion to aid removal of sulfur dioxide. Because walnut shells (and other biomass) do not contain a significant amount of sulfur, however, the gases can carry their heat into the boiler, resulting in higher efficiency. [California Energy Commission]

new insights to plants and to agriculture are for the most part still to come, but some examples of what the future may hold can already be found.

Standard agronomical techniques such as the selection of superior breeding stock have usually been applied to increase yields of food; selection to enhance total biomass production is just beginning. Experiments at the University of Georgia and elsewhere with short-rotation cropping (3 to 5 years) of sycamores and other rapidly growing species have shown that yields of wood can be as high as 40 tons per hectare—nearly three times the productivity of conventional forestry methods (Figure 54). Genetic manipulation techniques have given rise to hybrid grains and improved strains of microorganisms for fermentation, but these techniques are only beginning to be applied to trees and other species of interest for biomass energy. Screening of tree cells growing in culture to select new genotypes in the laboratory instead of by lengthy field trials is still in its infancy, and laboratory fusion of cells to form trees with hybrid vigor has not yet been accomplished. Examples of more advanced genetic engineering include attempts now under way to incorporate the ability to fix nitrogen into nonleguminous plants—an achievement that would reduce the need for energy-intensive nitrogen fertilizers—and to transfer the ability to ferment grain from yeast to more prolific bacteria.

Looking still further into the future are attempts in the United States, Japan, and Germany to transcend the plant itself and achieve photolysis of water (to produce hydrogen) in artificial photosynthetic systems composed of chlorophyll-containing elements extracted from plant cells. A second approach to the same end, espoused by Melvin Calvin of the University of California at Berkeley and others, is to construct artificial photochemical membranes that would, in effect, be the biochemical analogs of photovoltaic cells (see Chapter 3). These research efforts are in preliminary stages and are not expected to succeed overnight, but clearly the limits of the biomass resource are by no means fixed and unchangeable.

For the near term, practical interest is focused on relatively few crops and residues. Corn stalks, husks, and cobs in particular are regarded as readily available in quantities that could produce as much as 1 quad of energy in corn belt states, which now consume large quantities of propane and liquefied petroleum gas (LPG) to fire crop dryers and other farm equipment.

Sugarcane and sweet sorghum, as well as corn itself, might prove to be good energy crops (Figure 55). A study by Battelle Memorial Institute concludes that there might be a substantial near-term market for industrial alcohol (ethanol) fermented from these materials. About 300 million gallons a year of industrial alcohol are now made from ethylene, which in turn is made from natural gas or petroleum. Battelle's estimates are that biomass ethanol from a full-scale plant could sell for $1 to $1.25 per gallon, compared to about $1.15 per gallon for

Figure 54. A 7-year-old stand of tulip poplar growing on an upland site in Georgia as part of experiments in short-rotation forestry. The trees were planted in 4-foot rows with a 2-foot spacing between trees, as if they were a field crop such as corn. The young trees are easier to harvest with mechanical equipment and can be a source of either energy or high-grade pulp for making paper. The tulip poplars and other rapidly growing species cultivated in this manner yield up to three times as much wood as trees grown by conventional forestry techniques; in fact, harvesting the trees within 3 to 5 years appears to give the maximum yield. [Claud Brown, University of Georgia]

Figure 55. Test plot of sugar cane in Florida. New processing equipment now under development would slice the cane stalk and remove the sugar-containing pith, leaving the fibers of the stalk intact; the fiber can be used to make high-quality paper or plywood products. The sugar can be fermented to ethanol, and sugar producers in Florida and Hawaii are now eyeing industrial ethanol as an attractive second market. Sales of a valuable co-product such as cane fiber might be essential to lowering costs sufficiently that ethanol from sugar cane could be used as fuel. Experiments with other methods of improving yields and reducing costs are underway, including new strains, denser plantings, the use of solvent extraction to lower distillation costs, and methods of converting the cellulosic fibers of the stalk to sugar. [E. S. Lipinsky, Battelle Columbus Laboratories]

ethanol from ethylene. According to Edward Lipinsky of Battelle, the study director, further cost reductions in biomass ethanol would be possible if the processing equipment could be operated year-round and not just during the sugarcane harvest; for example, a mill could stockpile molasses, a sugar by-product, and ferment it when the cane harvest was over, or use sweet sorghum, a high-yield tropical grass that can be grown as a second crop in sugarcane areas. There is also growing interest in the idea that a second, energy-related market could help stabilize U.S. sugar and grain prices in years of plenty—a concept that has worked well in Brazil.

Some investigators have proposed that biomass ethanol can already be made cheaply enough to compete with gasoline as a motor fuel in some parts of the United States, although the claim is controversial. Large commercial facilities to produce ethanol for blending with gasoline are under consideration in Nebraska, where there is an excess of spoiled or poor-quality corn, and in Hawaii, where fuel prices are high and sugarcane is abundant.

Forest wastes in the United States are nearly as large as those from agriculture. They total about 24 million tons per year in unused mill wastes and 83 million tons left on the ground in the forests with current harvesting methods, according to a report by the MITRE Corporation. The report concludes that as much as 4.5 quads of energy per year could be produced on just 10 percent of now-idle forest and pasture land with wood grown in close-spaced, short-rotation tree farms using poplars, eucalyptus, or other high-yield species. Other investigators point to a huge unused reserve of wood—exceeding 10 quads—that is contained in stagnant stands of noncommercial species or diseased trees; if harvested, this biomass could have a major near-term local impact on energy supplies in New England or the Great Lakes region.

The forest products industry does not yet think of itself as a potential energy producer, and some industry spokesmen are skeptical that biomass in the quantity proposed in the MITRE report can be produced without cutting into the timber and paper markets. But a feasibility study done by the Forest Products Laboratory in Madison, Wisconsin, concludes that the U.S. forest industry could come close to being self-sufficient in energy by 1990 using wastes alone—a development that would expand present production from 1 to almost 3 quads of biomass energy.

A third major category of potential biomass resources is aquatic plants. William Oswald of the University of California at Berkeley has proposed growing blue-green algae on sewage wastes and has obtained yields of 16 to 32 tons of dry biomass per acre per year. Algae are readily converted to methane by anaerobic digestion, but they are also rich in protein and may lend themselves to the joint production of energy and food. Water hyacinth, a rapidly growing plant that now clogs many inland waterways, has also been proposed as a potential feedstock for methane production.

An even more ambitious proposal, by Howard Wilcox of the Naval Ocean Systems Center in La Jolla, California, is to grow huge rafts of kelp in the open ocean as sources of methane, animal feeds, and chemicals (Figure 56). To feed the kelp, nutrient-rich water would be pumped up from the lower levels of the ocean. The proposal has attracted the interest of the gas pipeline industry, which is helping fund preliminary research.

If adequate resources of the raw material can be found, there is a wide range of technologies for converting biomass to fuels or directly to energy, both under development and already in use. The list includes biological processes, aerobic and anaerobic fermentation, and thermal processes, including partial oxidation, pyrolysis, steam reforming, and direct combustion.

A technology of particular importance for dairy farm and feedlot use is biogasification with anaerobic fermentation. Animal wastes and some plant wastes are fed into a concrete or metal reactor, where they are decomposed by bacteria and converted to an intermediate-Btu gas composed primarily of methane. The residues from the reactor (or digester) are excellent fertilizers. In some countries, anaerobic fermentation is becoming an important energy source in rural areas. China is reported to have installed some 4 million digesters in recent years, and they are also being deployed in Taiwan and a half-dozen other countries. Biogasification is also being considered as a source of commercial energy production by a few companies in the feedlot and dairy industries in the United States and elsewhere. A development that may accelerate interest in the technology is an experimental reactor, developed by William Jewel at Cornell University, that has demonstrated gas production at ten times the rate of conventional digesters and with retention times as low as 3 hours.

Thermal means of gasifying biomass are also well established, especially partial oxidation, which has a history extending back before 1800. Gasifiers provide a simple means of retrofitting existing gas- or oil-fired boilers to burn biomass, as in the walnut factory in California. The product is a low-Btu gas that may require some derating of equipment, but conversion efficiencies of 80 percent and above are routinely achieved. Biomass in fact seems to be a nearly ideal feedstock for gasification processes because it is high in volatile constituents that are easily driven off at moderate temperatures. Experiments by Michael Antal of Princeton University with pyrolysis of cellulose have shown that 90 percent of the material is turned to gas at 450°C. By contrast, most coals heated to 900°C yield only 30 to 40 percent volatiles.

Most thermal conversion processes work best with relatively dry biomass, but steam reforming is equally applicable to wet biomass. A novel proposal by Antal is to couple steam refining with concentrating solar collectors (see Chapter

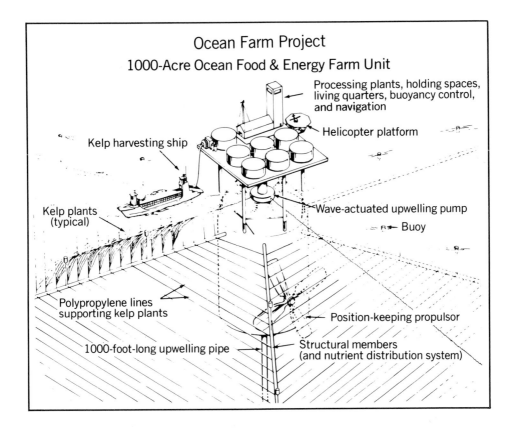

Figure 56. Artist's conception of an ocean seaweed farm that might produce large quantities of biomass for conversion to natural gas. Ocean farming has the advantage, of course, that it is not limited by the availability of either land or water. Nutrients are often in short supply in surface waters, however, and the concept calls for pumping nutrient-rich water from deeper layers to stimulate rapid growth of a particular seaweed, the giant kelp, attached to submerged lines. Methane would be produced from harvested kelp by anaerobic fermentation; animal feeds, fertilizers, and other by-products could be produced from the fermentation residues. Whether ocean farming of kelp for energy can be economically successful is still very uncertain. Preliminary experiments are underway off the California coast. [Howard Wilcox, Naval Ocean Systems Center]

Figure 57. A mobile tree harvester capable of shearing off an entire tree at the stump. The trees are then hauled to a central loading area with a tractor-like vehicle and mechanically loaded into a chip harvester that shreds the logs. The chip harvester feeds a stream of wood chips into a waiting truck. Woodsmen using this equipment can harvest about five times as much in a day as they could using chain saws. Wood chips can be burned directly. Several companies, however, are now building plants that make a more uniform and more easily handled fuel—small pellets made from wood chips or agricultural wastes that can be used interchangeably with coal. [Morbark Industries, Inc.]

4) to generate the steam. This solar-powered biomass refinery would produce hydrogen with an energy content 36 percent higher than that of the raw biomass and might be especially applicable to the manufacture of fertilizers and other chemical products. Antal estimates the efficiency of the combined solar-thermal and biomass process at higher than 70 percent, in part because steam reforming makes use of the moisture contained in most biomass rather than using a portion of the energy to drive it off first.

Similar processes exist for liquid fuels. Fermentation of sugar crops or grains to produce ethyl alcohol is not a new technology. Recent work at Purdue, however, has led to a new and efficient process for converting cellulose to sugars, which can then be fermented too, thus making possible much greater yields of alcohol. Solvent extraction techniques being developed at Battelle and elsewhere may reduce the energy used in distilling alcohol. Dry biomass such as wood can be converted to methyl alcohol by first gasifying it and then reacting the gas with a catalyst—the traditional process for making wood alcohol.

Even direct combustion of biomass, which must be the oldest energy technology known, is benefiting from some new and rediscovered technology. Major boilermakers in the United States now offer commercial systems for wood-firing boilers, either alone or in combination with other fuels. New equipment for

Figure 58. Pilot plant near Pittsburg, Kansas, for converting cellulose to sugar and then to ethanol. Cellulose is the structural material of all plants and makes up the bulk of nearly all agricultural, municipal, and forestry wastes. In the pilot plant, water is added to waste cellulose and the mixture is vigorously stirred to break up the fibers and form a slurry. The cellulose in the slurry is decomposed biologically with the aid of a mold, *Trichoderma viride,* that produces an enzyme known as cellulase; the enzyme catalyzes the breakdown of cellulose to glucose. Then other enzymes produced by yeasts promote the conversion of glucose to ethanol in a conventional fermentation process. Both reactions proceed simultaneously in the same reactors in the design being tested here. Finally, the alcohol is distilled from the mixture. From ethanol it is a short step chemically to ethylene, the basic building block of the chemical industry. Because of rising petroleum prices, ethylene is already more expensive than sugar in the United States, so chemical companies are becoming more and more interested in cellulose as a renewable feedstock. [Gulf Oil Chemicals Company]

pelletizing wood and thus making it a more uniform fuel for industrial applications is coming onto the marketplace, and the Georgia Institute of Technology is developing a transportable pyrolyzer that can produce a charcoal-like solid fuel from forest or crop wastes. Harvesting technologies for wood are also experiencing a wave of innovation (Figure 57).

In addition to direct production of energy, there is growing interest in the use of biomass as a chemical feedstock—a use that may eventually displace substantial quantities of fossil fuels that would otherwise be consumed. The production of fertilizer in anaerobic digesters has already been mentioned; the production of ethylene from ethanol and the production of furfural from wood are additional examples. Since chemical products generally have a higher value than energy, feedstock use may have some economic advantages over simple bioconversion. Major petrochemical firms are already looking into the potential of biomass feedstocks. A study by Dow Chemical, for example, identified four crop groups capable of supporting an industry equivalent to 1 billion pounds of ethylene: sugarcane, cellulosic wastes, corn, and sorghum. A subsidiary of Gulf Oil has been operating a pilot plant for converting cellulose to ethyl alcohol since 1976 and is designing a full-scale commercial facility (Figure 58).

Despite innovative ideas and claims of feasibility for a wide range of proposals, the economics of most biomass energy systems are still uncertain. Relatively few experimental efforts or field trials have been conducted, especially with crops grown for energy production, and most cost estimates are based on analytic studies or are extrapolated from outmoded equipment. These preliminary estimates are encouraging, however, and the competitiveness of biomass fuels seems certain to increase as supplies of oil and gas dwindle and costs rise. Synthetic fuels from coal now appear likely to cost between $4 and $5 per million Btu's, a target that many investigators believe can also be met with biomass systems capable of large energy yields.

One aspect of the Department of Energy program for synthetic fuels from biomass that concerns many biomass specialists is, paradoxically, its focus solely on energy. They point out that in the United States biomass production is so embedded in the food and fiber industries as to make multiple use of biomass resources the most economical approach. Lipinsky, for example, points to the potential of using corn as a feedstock for alcohol production, corn husks and stalks as fiber for paper production or fuel for grain drying, and the protein-rich stillage left over from alcohol fermentation as an animal feed—approaches that are already being pursued separately. There is as yet no experience with and no funding for integrated processing facilities—biomass refineries—and some critics doubt that the Department of Energy is institutionally capable of such an integrated approach. They note that it does not have a network of field stations capable of testing and tailoring multiple-use biomass processes to meet the various needs of farmers and foresters in different regions, and that the Department of Agriculture, which does have such a network, has no charter in the energy field.

Biomass is such an obvious and ubiquitous resource that its energy potential has been largely overlooked. New ideas for how to tap this potential are appearing at a rapid rate and many of them may be applicable in the near term. Before the United States is faced with the necessity of committing hundreds of billions of dollars to a fossil-based synthetic fuel industry, it would be advantageous to explore much more thoroughly whether biomass fuels can also play a major role.

7

WIND POWER
The Cheapest Solar Electricity Today

Perhaps because the first suggestions that emerged after the 1973 oil embargo for solving the energy crisis with wind sounded terribly grandiose—1000-foot-high windmills deployed across the Great Plains and the New England coast in staggering numbers—wind power has been unfairly branded as an offbeat energy source. Wind power also has to live with the unglamorous truth that it has been used for millennia by technically primitive cultures with rudimentary energy systems (Figure 59). Finally, like most other sources of energy that ultimately derive from the sun, wind bears the onus of being a fluctuating energy source.

Wind-energy research consequently receives markedly less support than other systems for producing electricity from the sun's energy, even though wind technology is far more advanced because of its long history of development. The new and relatively untested power tower concept for direct conversion of sunlight into electricity receives more than twice the support that wind research does, even though wind machines available today are at least three times cheaper than the first power tower pilot plant will be when it is completed in the 1980s. In fiscal 1978, wind power received only $35 million out of the $300 million solar research budget. Whether measured by the funding decisions of government energy research officials, the public support reflected in congressional lobbying, or the interest expressed by utility companies, wind power is lagging far behind.

Yet wind machines have been technically proven. They represent a kind of electromechanical technology that American industry can successfully mass-produce (Figure 60). Many types are commercially available and in selected locations they are already competitive. Without a doubt, wind energy is the cheapest form of solar electricity available today.

An important fact that is not widely appreciated is that in any reasonably windy region the amount of wind energy available on an annual basis is larger than the average energy flux of sunlight. The average wind power on the Great Plains over the course of a year is more than 200 watts per square meter. Two considerations could give wind energy the upper hand for applications in certain locations. First, the geography of the countryside provides natural concentrators for wind energy that may double its flux. In a low gap in the Rocky Mountains near Medicine Bow in southern Wyoming, for instance, the average annual wind speed is 21 miles per hour and the energy flux is 500 watts per square meter. Second, wind turbines are at least twice as effective as direct solar systems in extracting work from the natural medium. Wind-power systems routinely operate with 35 percent efficiency (the theoretical maximum is 59 percent), whereas solar systems that produce electricity achieve only 5 to 15 percent efficiency. Wind also has an inherent potential advantage in lower materials costs—the typical turbine blade covers only about 10 percent of the area from which energy is collected.

Numerous studies show that the deployment of wind machines in significant numbers would bring the costs down to the point where electricity could be produced for 2 or 3 cents per kilowatt-hour. Since this is about the cost of fuel oil, the market for wind is huge. Recent studies by SRI International and MITRE (Appendix 1) find that wind will become steadily more attractive, producing at least 2 quads of energy by 2000. This would be the equivalent of 80 large generating plants (80,000 megawatts). By 2020 wind could supply 4 to 9 quads—about 5 percent of the country's energy.

The major objections to wind installations have been raised against their aesthetic impact and possible interference with television reception. The visual assault posed by the gargantuan scale of early wind-power proposals would have been considerable, but the wind installations being built and projected for use by utilities are small by comparison. The tallest wind turbine planned will have a tower 200 feet high, and that will "probably be the maximum height" for a practical wind turbine, according to Lou Divone, who has headed the program under three successive agencies, the National Science Foundation, ERDA, and now the Department of Energy. A 200-foot tower would only be slightly larger than the towers for long-distance electric transmission, and perhaps less visible than the grain elevators that already dot the Great Plains. Television interference can be a problem for large wind machines, which usually operate at a constant

Figure 59. Valley in Crete where small water-pumping windmills help irrigate the fields. [Department of Energy]

speed not dissimilar from the television synchronization frequency. But the extent of the problem depends on the specific site. Large machines will cause interference for a range of 100 feet to 1 mile; for small machines interference is seldom a problem.

Until about 2 years ago, the government program concentrated overwhelmingly on large wind machines—those ranging up to 2.5 megawatts with rotors up to 300 feet in diameter. The agency's mission analysis studies, done by General Electric and Lockheed, showed that the cost of power from a wind turbine dropped rapidly with increasing size up to a power of 0.5 or 1 megawatt, and more slowly thereafter. From 1973 on, the program for large machines grew rapidly because "it was obvious where to put the money to cut costs," according to Divone. "All the available data said that scale factors favored large machines," he said. The government began an accelerated 5-year program to develop three large wind machines dubbed Mods 0, 1, and 2, rated to produce 0.1, 2.0, and 2.5 megawatts, respectively.

In the eyes of many observers, ERDA concentrated on large wind turbines partly because agency officials thought that anything less would have a negligible impact on total U.S. energy supplies. But a thorough assessment of the sizes of the potential markets for large and small wind machines has yet to be made.

Over 6 million windmills have been built and bought in the United States since the middle 19th century, and modern-day applications for heating and cooling, water pumping, and dispersed electricity generation could see a resurgence. "The market for small-scale machines," said a respected dean of the wind energy community, Frank Eldridge, "might be of the same order of magnitude in terms of total power as the market for large-scale machines in centralized electric power applications."

After a few months of testing, it became clear that the first large machine to be built, Mod 0, had run into trouble. The 100-kilowatt machine, built by NASA at its Plum Brook field station 50 miles west of Cleveland (Figure 61), developed severe forced oscillations and unexpected impulse loads on the propeller. The problems were so severe that it logged only 57 hours in the first 8 months of operation. ERDA officials were afraid for a while that the machine might throw off a blade because of metal fatigue caused by the unexpected stress. The problems with the Plum Brook machine have been fixed now, but its usefulness as a test-bed for advanced components was severely curtailed by the delays incurred. The problems at Plum Brook may also have influenced the agency's decision to cancel long-standing plans to test fiber glass blades on the Mod 0 and Mod 1 machines. The Mod 0 was largely inspired by a German wind machine designed by U. Hutter that operated successfully for 10 years with fiber glass blades.

The program has proceeded by effectively sidestepping the near failure of the initial wind machine (which ultimately cost $2 million), and the energy agency is building three upgraded versions of the Mod 0 for regions of higher wind velocity, using a 200-kilowatt generator on the same-sized turbine. However, 6 months after the first upgraded Mod 0 was installed at Clayton, New Mexico (Figure 4), it also ran into trouble. Engineering around the stresses on a large machine again posed a problem. Cracks and missing rivets constituted enough of a problem that NASA removed the propeller blades for repairs. The next step, the Mod 1, will have a much larger rotor, 200 feet in diameter, and will be designed to produce 1.5 to 2.0 megawatts in high-wind regions. The largest in the series, Mod 2, will have a rotor 300 feet in diameter. It is due to start turning in late 1979 and will cost $10 million.

Figure 60. Multibladed windmills were widely used to pump water on American farms before rural electrification. Early in this century, the mass-produced American farm windmill sold for about one-fifth of what it would cost today, after correcting for inflation. Over 6 million of these windmills were built before their use declined in the 1930s.

[Department of Energy]

Figure 61. The first large wind turbine built by the federal program, Mod 0, located near Sandusky, Ohio. Blade "shadow" from the stairs posed problems when the 100-kilowatt machine started operation, and the stairs were eventually removed. [Department of Energy]

The large-scale wind turbine program is a prime example of a sequential development philosophy. Each of the series of wind turbines is a larger version of the same concept, namely a horizontal-axis wind turbine with a two-bladed rotor oriented downwind from the tower. The upgraded versions of the Mod 0 are being built by Westinghouse and Lockheed. The Mod 1 is being built by General Electric and the Mod 2 by Boeing. A program with a parallel development philosophy would have tested other concepts, such as three-bladed rotors, upwind designs, and vertical-axis turbines, to get a broader basis for assessing costs and reliability before proceeding to larger sizes (Figure 62). The danger of the narrow sequential approach being taken by the Department of Energy is that the considerable potential of large wind machines could be lost if it turns out that the wrong design was chosen at the start of the sequence.

Before the government could begin operation of a 200-kilowatt, upgraded Mod 0 wind machine destined for Block Island, a private company built a similar-sized wind machine—reportedly for much less money—on another island not far away. The inhabitants of Cuttyhunk Island, between the Massachusetts mainland and Martha's Vineyard, wanted a windmill to offset the skyrocketing cost of fuel for their diesel generating system. A company named WTG, Inc., in Angola, New York, wanted a site to test a prototype for a large wind system. Because the uncle of one of WTG's founders had a summer home on Cuttyhunk, the two parties got together. A 175-kilowatt wind turbine was installed on the island just outside the town of Gosnold in the summer of 1977 (Figure 63) and linked with the existing diesel unit to form a hybrid system. The turbine reaches its maximum rated power at a wind speed of 28 miles per hour, and the average annual wind speed on the island is 16.8 miles per hour.

For reliability, the WTG rotor was built to be as simple as possible. It is a fixed-pitch, three-bladed steel rotor, 80 feet in diameter, on an 80-foot tower. The system was modeled after the Danish Gedser mill in many respects, according to Alan Spalding at WTG. The Cuttyhunk machine, built by a ten-person private company, reportedly cost $280,000. Spalding says the factory price of subsequent machines, not including shipping or installation, would be $210,000. The neighboring Block Island machine, built by ERDA, NASA, and aerospace contractors, cost $2 million.

Another large wind machine has been privately developed by Charles Schachle, a Seattle inventor who worked for Curtiss-Wright aircraft company in the 1930s and began designing a large modern wind turbine in 1970, well before the surge of government interest in wind power. The Schachle wind turbine generator is conceptually different from the government-sponsored wind turbines in many respects. The machine uses a three-bladed rotor that turns at a variable speed depending on the wind velocity rather than at constant speed like the large machines in the federally sponsored program. "Riding the wind" in this fashion

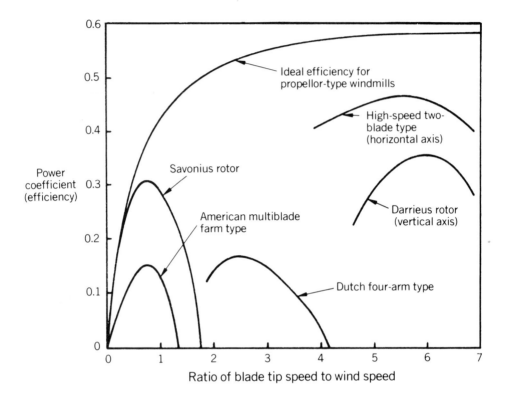

Figure 62. Typical performances for different classes of wind machines. A common measure of wind-machine efficiency is the ratio of the blade power of the machine to the energy of the wind moving through the blade area. This efficiency has a well-defined peak for each type of machine, depending on the ratio of bladetip speed to wind speed. The theoretical maximum performance is about 59 percent. The various types can be divided according to whether they employ lift or drag to obtain power. The American farm multiblade windmill is an example of a machine which has low efficiency and turns no faster than the wind. The advantages of the design are its high torque at low speeds, which is ideal for water pumping, and the fact that the blades can be any convenient high-drag shape and are therefore easy to manufacture. The Dutch four-arm windmill is a lift-type machine that has inefficient performance because of the crude design of its sail-like rotor, but it also produces high torque. The Savonius rotor, which looks like a vertical cylinder with the two halves displaced in opposite directions from the axis, is a machine that employs a combination of drag and lift. High-performance modern wind machines using lift-type blades with carefully designed airfoil shapes perform best when they are moving considerably faster than the wind. The horizontal-axis propeller-type wind machine has a maximum efficiency of about 47 percent, which appears to be the highest achievable without special tricks. The vertical-axis Darrieus rotor, also a lift-type machine, is less efficient than the propeller type but is thought to be a promising candidate for economical power production because of the simplicity of the design and the virtue that it is insensitive to the direction of the wind. [Robert E. Wilson, Oregon State University]

enables the machine to maintain the optimum ratio of blade tip speed to wind speed and thus maximize its efficiency through a range of wind velocities. Another significant difference is that, to accommodate the variable speeds, the rotor is coupled to the generator through a hydraulic system rather than a mechanical gear linkage. Another distinction is that the entire tower rather than just the rotor assembly turns in order to keep the rotor faced into the wind. Finally, the blades are made of laminated wood with a fiberglass leading edge, rather than metal.

Since May 1977, Schachle has tested a 72-foot-diameter version of his design at Moses Lake, Washington (Figure 64). Working with his sons, who have degrees in electrical and aeronautical engineering, Schachle designed the machine in seclusion in a rented aircraft hangar. He says he has put "a few million" of his own money into the prototype, but not a cent of government money.

Southern California Edison Company spent about a year looking at different large wind turbine designs for installation at a site of high annual winds near Palm Springs. According to the technical advisor at Southern California Edison, Bob Scheffler, the company evaluated all the large turbines, including those in the Department of Energy program, before deciding to buy the Schachle machine. The upgraded Mod 0 was more expensive, and the Mod 1 was undergoing many revisions of design and costs were therefore uncertain. Furthermore, the year of experience that had been compiled with the 72-foot prototype put the design ahead of both the government machines and the private machine on Cuttyhunk, according to Scheffler.

Southern California Edison contracted with Schachle, whose company is named Wind Power Products, Inc., in May 1978 to build a 165 foot diameter machine for its test program at a cost of $1 million. Southern California Edison estimates that subsequent turbines might cost half as much and, if so, would produce power at 3 cents per kilowatt-hour.

A large-scale test of a vertical-axis wind turbine is also under way (Figure 65) under the sponsorship of the National Research Council of Canada. The device is a two-bladed Darrieus machine located on the Magdalen Islands in the Gulf of St. Lawrence. Erected in May 1977, the 200-kilowatt wind turbine supplies power to the island grid and is backed up by the islands' diesel generating plant in a manner similar to the Cuttyhunk system. At the end of the first year, the machine had operated intermittently, according to R. J. Templin at the Research Council, and reached a maximum power of 120 kilowatts with "lots of teething troubles." In mid-1978, the spoilers on the machine failed during tests, causing the blades to accelerate and break apart.

The Darrieus design is generally considered to have the potential to be cheaper than horizontal-axis machines, if the particular structural problems associated with the design can be solved, because the blades can be thin and

Figure 63. A privately financed 175-kilowatt turbine installed on the island of Cuttyhunk, Massachusetts, in summer 1977. The machine was designed by WTG, Inc., of Angola, New York, and is located near the town of Gosnold. The illustration shows the machine prior to installation of the cabin housing. [WTG, Inc.]

light. The structural problems arise because the blades on a Darrieus undergo fluctuating torques, which drop to zero on each blade as it passes through the eye of the wind. But there is no need for a yaw mechanism to keep the blades into the wind, and the heavy equipment—such as the generator—can be located on the ground. The 200-kilowatt Darrieus machine was manufactured by Dominion Aluminum Fabricating, Ltd., Toronto. Dominion and another Canadian company, Bristol Aerospace Ltd. in Winnipeg, also manufacture smaller vertical-axis wind turbines.

While the paper studies that inaugurated the wind program all showed that large machines could be made cheaper per kilowatt than small machines, practical experience has yet to bear this out. The machines built for the government have been particularly expensive. The Mod 0 at Plum Brook cost more than $5500 per kilowatt. The Mod 1 machine, due to be installed at Boone, North Carolina, is projected to cost about $2000 per kilowatt, but some features were changed late in the design and General Electric has requested substantial cost overrun compensation. The Mod 2 is projected to cost about $1800 per kilowatt. Each of these machines is a prototype for its scale, and the more important question is what the same machine would cost in production. A survey of the question by JBF Scientific, drawing on all the economic studies of large machines made so far,

Figure 64. A 72-foot-diameter prototype of the Schachle wind turbine generator, located at Moses Lake, Washington. The turbine, which is rated at 140 kilowatts in a 26-mile-per-hour wind, was erected in May 1977. The turbine is fixed on the tower, and the whole tower pivots about a post at the bottom in response to signals from a wind indicator in order to continually face into the wind. Before building this prototype, inventor Charles Schachle tested propellers on a 24-foot scale. He has contracted to build a 165-foot-diameter machine for Southern California Edison, at a cost of $1 million, for installation by April 1979. The large wind turbine will be located near the San Gorgonio pass in Southern California, not far from Palm Springs. Westerly winds blow through the pass at an annual average speed of 18 miles per hour. The San Gorgonio wind turbine will reach its rated maximum of 3 megawatts at winds of 40 miles per hour. Above 40 miles per hour, the blades will be feathered to avoid overloading the generator. Southern California Edison plans to test the large turbine for 1 or 2 years to evaluate its power production in the utility grid. It is estimated that the windy area near the pass is large enough to accommodate several hundred such wind turbines. Southern California Edison is the United States' fourth-largest electric utility. Because half of the company's yearly output is produced from oil, there is ample opportunity to use wind systems to save fuel. [Charles Schachle]

found that the cost would drop to $400 to $900 per kilowatt after the first 100 units were built. Divone, speaking for the wind program, said that the cost the agency realistically hopes to achieve is $750 to $1000 per kilowatt. By comparison, the Department of Energy's small wind program projection is that a cost of $500 per kilowatt should be an attainable goal, and small machines that cost $2000 and less per kilowatt are available today.

How much cost reduction can actually be achieved in building the large machines—particularly under aerospace direction—is an open question. Economic studies made so far assumed that the production experience would follow a "learning curve," whereby costs would drop 5, 10, or 15 percent each time production doubled, but no two studies could agree on the rate of cost decline. A more detailed approach would be to look at the likely mature product costs of the individual components. This approach was only beginning in summer 1978. But to appreciate what league large windmills are in, it is useful to know that the blades for the Mod 1 machine will be almost identical in size to the wing of a 747 aircraft, and the blades for the Mod 2 will be the size of the wings on Howard Hughes' famous "Spruce Goose." Economies of production may be more applicable to wind machines the size of helicopters and automobiles—technologies that have been successfully mass-produced in the past—rather than giant aircraft.

Even if experience shows that there are some economies of scale to be realized from the large machines, critics question whether space agency management and aerospace company engineering is the right recipe for progress. The technical management for the Mod 0, 1, 2 series has been handled by NASA, out of its Lewis Research Center, and the machines have been designed with high technology and aerospace materials. But many observers think that careful cost management is what is needed to make the big machines competitive, and this is a skill at which the aerospace industry hardly excels. A shift away from the aerospace development philosophy and toward a more aggressive attack on the key problems—which are cost and reliability rather than performance—is needed to determine the cost potential of large machines.

In mid-1976, after it was clear that the large-wind-machine program had gotten off to a shaky start, Divone's office, which was then part of ERDA, set up a small-wind-machine program and gave the job of technical management to the Rocky Flats laboratory near Golden, Colorado (Figure 66). (Although the principal mission of the laboratory is the fabrication of nuclear weapons parts, Rocky Flats is in a very windy location just east of the Front Range of the Rocky Mountains.) The small wind program received only $2 million in fiscal 1977, but its budget was boosted to $8 million in fiscal 1978 and was increased to $15 million in 1979. Because it started earlier, the small wind program is better developed than most other federal programs for small-scale solar technologies, except perhaps solar heating and cooling. But it still lags far behind the large wind

Figure 65. A 200-kilowatt Darrieus wind turbine located on the Magdalen Islands in the Gulf of St. Lawrence. The vertical-axis wind machine is 120 feet tall and 80 feet in diameter. The blade is a thin 2-foot-wide airfoil. Visible at the left are spoilers, which are used to keep the blade speed constant at high wind speeds. Two-bladed Darrieus wind turbines are not self-starting, so the 200-kilowatt generator also serves as the starting motor. Two rather than three blades were chosen—in spite of the fact that the aerodynamic loading is more troublesome for two blades—because the two-bladed design reduced manufacturing costs and made the turbine easier to erect. The turbine, funded by the National Research Council of Canada and operated by the provincial electric utility, Hydro Quebec, is believed to be the largest vertical-axis machine in existence. The wind turbine starts up at a wind speed of 12 miles per hour and reaches its rated power at a speed of 30 miles per hour. The average annual wind speed is about 19 miles per hour, and the turbine is expected to produce 600,000 kilowatt-hours of energy per year when fully operational. [National Research Council of Canada]

Figure 66. The small wind machine test site at the Department of Energy's Rocky Flats plant near Golden, Colorado. Located near the Front Range of the Rocky Mountains, the site encounters some of the highest winds in the contiguous 48 states. The site has towers for testing up to a dozen small wind machines. Seen here is a 25-foot-diameter wind turbine manufactured by the Grumman Corporation. The design is a three-bladed downwind turbine, rated to produce 15 kilowatts of power in 26-mile-per-hour winds. [Department of Energy]

program in research expenditures. The Rocky Flats program is intended to test existing machines and to develop new ones, ranging in size from 1 to 60 kilowatts or higher.

Several dozen small wind machines made by about 20 companies are already on the market (Figures 67–72). One of the most successful machines from the 1930s, the Jacobs Wind Electric plant, is being reconditioned and sold for $1100 to $1800 per kilowatt by Northwind Power, a small company in Warren, Vermont. Sized at 2 and 3 kilowatts, the Jacobs machines were sturdily built to provide reliable power. The size is appropriate for lighting remote cabins and waterway marker buoys. Some may have also been used to resist galvanic corrosion along the Alaska pipeline. Working independently of the government, the Grumman Corporation has developed a larger wind turbine, rated at 15 kilowatts (in a 26-mile-per-hour wind), that is a suitable size for supplying the electricity needs (except electric heating) for homes and farms. Grumman considers the machine a prototype rather than a commercial model, but at least a dozen have been sold for about $1300 per kilowatt. A small company in Lowell, Massachusetts, U.S. Windpower Associates, is developing a still larger machine, which would produce 25 to 30 kilowatts. No machine is available yet in a size approaching 40 kilowatts, but there is expected to be a substantial market for such a size for isolated working camps, small communities, and small factories. Many of the small machines are foreign-made, and one of the most reliable has been the Australian Dunlite (Figure 73).

Although the present small machines are handmade, they are remarkably close to being competitive with the price of electricity in some areas. Studies made by wind-power enthusiasts conclude that the present machines can produce electricity at 5 cents per kilowatt-hour in windy areas. The price of electricity from the utilities ranges from 2 to 10 cents per kilowatt-hour in different parts of the country, so the claims must be considered seriously. A study prepared by JBF Scientific Corporation for the Department of Energy's wind program is more pessimistic, but implicitly lends some support to the 5 cent per kilowatt-hour claim.

The JBF study concluded that a 10-kilowatt wind machine available today would produce electricity for 15 cents per kilowatt-hour over its lifetime, if it were used as a home system integrated into the utility grid without storage. If electricity prices were projected to rise 6 percent annually, then JBF concluded that the same wind system would be competitive today in areas where the electricity costs are over 8.7 cents per kilowatt-hour. But the JBF study was rather conservative because it assumed only 12-mile-per-hour average annual winds, whereas higher values are available in many regions. It also assumed no federal incentives, even though the Energy Tax Act of 1977 (or some similar provision) seems likely to give wind energy a tax credit. In a region with 15- to 18-

(67)

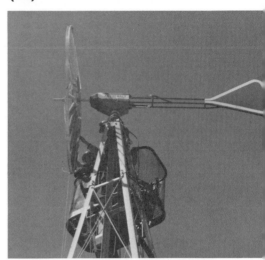

(68)

(69)

(70)

(71)

Figures 67–71.   The Jacobs Wind Electric plant is an old and well-proven small wind turbine. Over 80 reconditioned units have been sold in the past several years. Shown here **(67)** is a model that produces 3 kilowatts of electricity. It is 14 feet in diameter. The bicycle-turbine design **(68)** is reminiscent of the farm windmill, in that it has fixed blades oriented like the spokes on a bicycle wheel, but it gets somewhat higher efficiencies. The Amerenalt machine shown is 8 feet in diameter and produces 1.5 kilowatts of electricity. The Sencenbaugh machine **(69)** is designed for light-duty generation in remote areas. It has fixed-pitch wooden blades and generates 12-volt direct current. The machine is rated at 3/4 kilowatt. The Swiss company, Elektro, manufactures a range of small wind turbines, which have variable-pitch blades. Shown here **(70)** is a 6-kilowatt model that is 16 feet in diameter. One of the largest small wind machines available is the French Aerowatt **(71)**. This unit, with a 31-foot blade diameter, is conservatively rated to produce 4.1 kilowatts. Wind machines undergo severe stresses in high winds, and the unit shown here was destroyed in early 1977 when a 110-mile-per-hour wind swept across the Rocky Flats test site. Experience is proving that the Rocky Flats site is a severe test of any design, because it receives destructively high winds several times a year. [Rocky Flats Small Wind Systems Test Center]

Figure 72. One of the most powerful American-built small machines is the Zephyr, built by a company of the same name in Brunswick, Maine. With a 15-foot blade diameter, the Zephyr is rated at 15 kilowatts and is commercially available today. Like most small machines, the Zephyr is self-tending, in that it has provisions to limit the blade speed in high winds to protect the machine. [Rocky Flats Small Wind Systems Test Center]

Figure 73. A vintage 1930 small wind turbine that is still popular, the Dunlite machine is made in Australia. Its 12-foot rotor produces 2 kilowatts in a 25-mile-per-hour wind. [Rocky Flats Small Wind Systems Test Center]

mile-per-hour annual winds, the wind speed consideration alone would reduce JBF's calculated cost by a factor of 2 or 3.

Although the small machines are commercially available and seem economically attractive, as of 1978 the federal government has no small wind demonstration program. (It has spent hundreds of millions of dollars for solar heating and cooling demonstrations.) The Rocky Flats office has, however, begun government-funded competitions for the construction of three types of small machines, 1, 8, and 40 kilowatts in size (Figures 74–80). The burgeoning small wind program is a "pretty good slug of money for a little industry that had nothing 15 months ago," says Richard Katzenberg of the American Wind Energy Association, who, like many of the organization's members, is a small-wind-machine builder himself. Although wind had a notably weaker lobbying force than solar energy during the early 1970s, the American Wind Energy Association has become quite effective and it successfully lobbied for a $20 million increase in the $40 million wind budget for fiscal 1979.

The wind energy program in the United States is still in its infancy. The large wind program, although 7 years old and growing, is on a very narrow track. Small wind machines are thriving much more through the efforts of private individuals, often with very limited capital, than the federal program. Little or no analysis has been made of the market and price-reduction potential of small machines, and the crucial issue that will determine ultimate costs of different-scale machines—namely the trade-off between economies of scale for large machines and economies of mass production for small ones—have not been addressed. Furthermore, little attention has been paid to ways to accomplish heating from wind power, even though heating might be a more efficient use than electricity generation. The seasonal profile of wind energy is better suited to space heating than is the profile of solar energy, especially in the northern United States. The wind program, under Divone, is generally credited with good technical management, given the sequential nature of the program and its aerospace constituency, and with flexibility and attentiveness to all segments of the wind community. The very small staff—no more than four professionals since its inception—has been a severe bottleneck, however, which at times has slowed the limited supply of research funds to a trickle.

Wind power does not appear to be an example in which early emphasis on large-scale machines was completely mistaken. Small-scale systems may be ideal for semirural and rural installations, but a 30-foot-diameter windmill on every house is hardly a realistic ideal for a high-density urban environment. Although the optimum size of the larger machines is an open question, it appears that both large and small scales have economic potential. In fact, large and small systems are competing. But wind-power research has been badly underfunded, and the

**United Technologies
Research Center (74)**

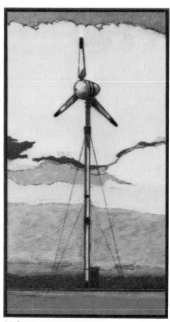

Windworks (75)

Grumman (76)

Alcoa (77)

Figures 74-77. Small wind machines are being sponsored by the small-scale wind program of the Department of Energy in a competition to determine the cheapest design in each of three sizes, rated at 1, 8, and 40 kilowatts. The winning machine in each category could be produced in quantity, to assess cost and performance and to use in government demonstration programs. Illustrated here are the designs in the 8-kilowatt category, which is intended to provide for the demand of the average home (without electric heating). Each of the 8-kilowatt machines has a rotor about 32 feet in diameter, whether the rotor turns on a horizontal or a vertical axis. The horizontal-axis machines do not require a tail to orient the rotor properly, since they are pivoted so that the rotor stabilizes in a position downwind from the tower. United Technologies Research Center, in East Hartford, Connecticut, has developed a two-bladed downwind machine **(74)** with a self-adjusting hingeless blade, which makes use of a flexible graphite-epoxy material so that the blade will flex to reduce its load in high winds. The design also uses an induction generator, which is attractive for cost saving. The Windworks design **(75)**, by a small company in Mukwonago, Wisconsin, will have three blades downwind and will use a direct drive from the propeller to an alternator to save money by avoiding the use of a gearbox. As in the United Technologies design, the blades on the Windworks machine will be hinged to flex (or "cone") in high winds. The Grumman design **(76)** will be a larger variation of the 25-foot machine already for sale by Grumman Energy Systems, Inc., Ronkonkoma, New York. It will be 32 feet in diameter (designed for lower wind speeds than a previous Grumman model) and will also be a three-bladed downwind design. It will use aluminum blades, a gearbox, and an alternator. The Alcoa entry **(77)**, designed by Alcoa Laboratory at Alcoa Center, Pennsylvania, is an outgrowth of research at Sandia Laboratories on the Darrieus machine for which Alcoa fabricated the blades. The Alcoa machine will be set on a fairly high tower for a Darrieus design (the base of the rotor will be 40 feet above the ground to take advantage of higher wind speeds at that height). The three-bladed rotor will be connected to an induction motor at the base of the tower, which will serve as both starting motor and generator. The configuration with the generator at ground level allows for easier maintenance. The cost goal for these four designs is $750 per kilowatt, assuming a production run of 1000 units. [Rocky Flats Small Wind Systems Test Center]

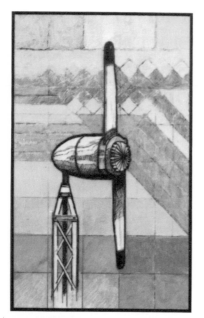

Enertech (78)

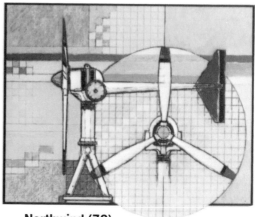

Northwind (79)

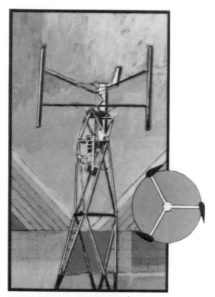

**Aerospace Systems /
Pinson Energy (80)**

Figures 78–80. Three wind machines being designed under a Department of Energy competition for a small 1-kilowatt wind machine, of high reliability, primarily for use in isolated areas. The Enertech design (**78**) is a two-bladed downwind design built by a small company in Norwich, Vermont. A three-bladed upwind rotor (**79**) is being developed by Northwind Power Co. of Warren, Vermont. The vertical-axis entry in the 1-kilowatt competition is the "cyclo turbine" design (**80**) by Aerospace Systems, Inc./Pinson Energy Corp. of Burlington, Massachusetts. It has three vertical blades that change pitch as they rotate to present the optimum angle of attack to the wind (unlike the Darrieus which has constant pitch). The cost goal for these three machines is $1500 per kilowatt, assuming 1000 units. A competition is also planned for a 40-kilowatt machine, with a cost goal of $500 per kilowatt. [Rocky Flats Small Wind Systems Test Center]

crucial questions of the best design philosophy and manufacturing philosophy have been woefully neglected.

Wind power does not need a breakthrough to put it in the right economic ball park to be competitive. Small systems may already be attractive, and a twofold price reduction would make wind systems competitive in many regions of the country. Since only 10 to 20 percent of the cost of present wind machines goes into materials, the opportunities for cost cutting—and the concomitant potential for an expanding wind energy market—are great. Wind power ranks at or near the top in projections of the contributions various solar energy technologies can make after the turn of the century.

ENERGY STORAGE
An Overstressed Problem

What do you do when the sun goes down? is a question often asked about solar energy. During cloudy days, long nights and times when the wind ceases to blow, what is to be done for power? The answer, according to conventional wisdom and seemingly impeccable logic, is to build an auxiliary system that will store energy when the sun is out or the wind is blowing and use the stored energy when they are not. The problem with the answer is that storage systems available today tend to be cumbersome, costly, and less than perfectly efficient. When their expense is added to the cost of a basic solar energy or wind system, which is no more than marginally economic in most regions, the total price often becomes exorbitant.

For this reason, the problem of energy storage is often characterized as a major obstacle to the widespread use of solar energy. The magnitude of the perceived problem has heavily influenced the public debate over the role that solar energy can play. It has been used as an argument for holding down solar research support, on the grounds that the cost of storage limits the potential use that can be made of solar energy for the indefinite future. It has been cited by government administrators as a prominent reason for giving strong support to certain solar technologies—not intermittent in nature—which might otherwise be difficult to justify. Even the prospect that the storage problem might be solved has been turned against solar energy programs, with the argument that a technical breakthrough that produced cheap storage would be more to the advantage of non-solar technologies than solar ones.

These arguments could be valid under certain assumptions, but the conclusions are nevertheless open to challenge. Solar energy storage is a complex problem, and the importance of storage to solar energy—especially for uses in the near- and intermediate-term future—may be overstressed and misunderstood.

One reason that storage springs to mind when solar energy is mentioned is that solar energy is often associated only with technologies that translate sunlight directly into thermal energy or electricity. In contrast to a few years ago when heating and cooling of buildings was its cutting edge, solar technology today includes a broad spectrum of concepts that differ greatly in their requirements for storage. Plant matter, or biomass, is ideally storable for long periods. Wind power is another form of indirectly derived solar energy which, even though it is intermittent, is available more of the time—day and night—than direct sunlight in many regions. Energy storage is more nearly indispensable for solar water heating, heating of buildings, and production of heat for industrial processes. But even in these applications there may be natural solutions, such as passive heating and cooling of homes in which normal building materials serve the storage function, that do not require any storage systems as such. There may also be the possibility of economic breakthroughs in thermal storage, possibly via annual storage on a community-wide scale, that could reduce costs and dramatically improve the reliability of solar heating.

Another factor that may work to subtly reduce the storage problem is any movement in the energy economy toward greater coordination of the sources of energy supply and demand. It would be naive to suggest that people will easily give up the prerogative of energy when and where they need it, but the flexibility of use that has occurred in an era of cheap energy may be changing. Time-of-day electric pricing, already introduced in some areas, is one evidence of this. The utilization of computer systems that automatically manage the energy load in large office buildings is another. Such changes, although they may be exceedingly gradual, will tend to make the alternatives to solar energy a little more intermittent themselves and may create a social climate in which solar energy may become more acceptable used "as is" rather than with storage.

The major reason, however, that storage may not be a solar sine qua non for the immediate future is that there are no fewer than three configurations in which solar power sources can be integrated with the present energy systems—particularly electric systems—so that new bulk storage is not required.

Overlooking these alternatives has led to a confusion of solar energy's near-term and long-term needs for storage. Even at wildly optimistic rates of growth, solar energy's contribution will be a relatively small change in the national energy

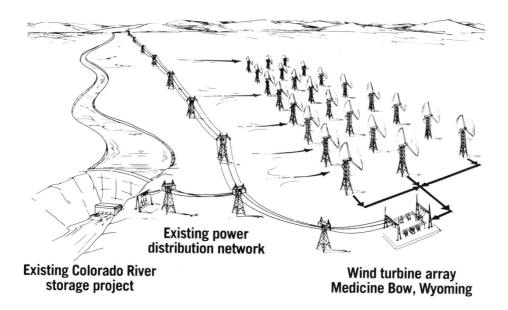

**Existing power
distribution network**

**Existing Colorado River
storage project**

**Wind turbine array
Medicine Bow, Wyoming**

picture until the turn of the century. In such circumstances, the traditional energy system can be used to compensate for the fluctuations of solar sources. So long as oil is a major component of the energy supply and is wastefully used where other fuels would suffice, solar energy systems can profitably be used to displace oil when the sun is out and left idle when it is not. Solar energy could also be used to match fluctuating energy demand, particularly for electricity, without endangering the stability of the networks. The national energy system is a well-integrated complex. Since large blocks of electric power are routinely transmitted long distances, the electric network is particularly well-suited for smoothing and balancing solar-source power fluctuations. If the solar-derived energy grew to too large a fraction of the total, the overall stability of an electric network might be adversely affected. There are, however, a number of studies which indicate that this limitation is unlikely to be a problem until the solar power penetration reaches 15 or 20 percent.

A strategy based on hybrid systems without storage would appear well suited for the upcoming decades. In that period solar energy may grow, but must nevertheless remain less than about 25 percent of the total energy mix because the turnover in systems and equipment is too slow for more rapid change to occur. It is only when the supply of oil begins to dwindle and the penetration of

Figure 81. A proposed system that would produce electricity from wind with 100 percent reliability by linking a large field of wind turbines in Wyoming with existing hydroelectric plants in the Colorado River regional system. The field of wind turbines would be located in a lightly populated area near a low gap in the Rocky Mountains near Medicine Bow, Wyoming. With an average annual wind speed of 21 miles per hour, the site has the highest winds measured in the state. The amount of power available peaks in midwinter, reaching a maximum of about 800 watts per square meter in late February and dropping to a low of 200 watts per square meter in early August. Long-distance transmission lines link the wind turbine site with the Flaming Gorge Dam in northeastern Utah and the Glen Canyon Dam in northern Arizona. The effect of storage would be achieved by holding back water when the wind was blowing. Together, the dams have sufficient excess capacity to match the power (98 megawatts) that would be produced by the proposed wind system. The Medicine Bow site has such favorable winds (12- to 18-mile-per-hour average winds are often considered excellent) that it has been estimated that power could be produced for 2 cents per kilowatt-hour, assuming the 49 turbines in the system would be of the Mod 1 design (see Chapter 7), initially costing $1300 per kilowatt and dropping rapidly to $500 per kilowatt by the time the 49th unit was built. [Stanley Hightower and Abner Watts, Bureau of Reclamation, Denver]

solar electric technologies into the national electric grid becomes more substantial that cheap energy storage will be badly needed.

Thus solar energy's long-term problem has been somewhat unfairly portrayed as an obstacle to its development in the near term, when in fact there are many options. It is particularly ironic that some of the critics who say that solar energy cannot supply more than 1 or 2 percent of the country's energy by 2000 also cite storage as a critical problem. (At that low a growth rate, storage would only be a problem for the 22nd century.) The danger in focusing on the wrong time horizon is that solar energy may be developed at such a leisurely pace that it will make less contribution than it could to a mixed energy economy.

The first alternative to building expensive new storage systems in the near term is to use existing storage systems, those of hydroelectric installations. By holding back water that would otherwise be flowing out of a hydroelectric dam, energy can be stored in one part of an electric network while a solar energy system (perhaps a photovoltaic or wind system) is producing energy in another part. The United States has 59,000 megawatts of hydroelectric capacity, and an additional 10,000 megawatts of pumped storage capacity. Pumped storage facilities consist of a pair of reservoirs, the upper one being filled with water pumped up from lower levels at times of minimal electric usage. Although half the hydroelectric capacity is in the three West Coast states (the largest single project is the 4000-megawatt Grand Coulee Dam on the Columbia River in Washington State), hydroelectric dams are found in 47 states. New York State has a particularly large resource, with an extensive network of small dams in addition to the one at Niagara Falls.

Little alteration, if any, is needed to utilize hydroelectric projects for storage. In instances where the plant was already being used at full capacity, the installation of extra turbines ("overmachining") would be needed. But most hydroelectric facilities regularly run at much less than full capacity (Grand Coulee generally uses only about half of its 21 turbines), so there would be extra capacity available to produce a surge of power when a wind or solar system was down. Although this is not literally energy storage, it has the same effect as if the wind or solar energy had been used to pump extra water into the hydroelectric system's reservoir for release later.

A prominent proposal to use this sort of system in the Rocky Mountain region has been made by Stanley Hightower, at the Bureau of Reclamation in Denver. After analyzing the extra capacity available from the dams in the Colorado region, particularly the Flaming Gorge and Glen Canyon dams, Hightower and Abner Watts concluded that the available hydroelectric storage capacity was sufficient for a 98-megawatt system of large wind turbines. Making use of the flexibility that the electric network offers, Hightower suggested that the wind turbines be located in a site of particularly high annual wind speeds, near Medicine Bow, Wyoming (Figures 81 and 82).

A different sort of proposal for a new wind, old hydroelectric project is under consideration in New York State by the Niagara Mohawk Power Corporation. In that instance, the wind system would not be hooked directly into the electric grid. Situated near a small hydroelectric plant on the Salmon River east of Lake Ontario, the proposed Niagara Mohawk wind-hydro system would pump water back up into the reservoir for use again and again. In that region, the winds and the electricity demands both peak in the winter, when the water levels are low. By recycling water, the storage necessary to make wind a reliable power source can be achieved, according to S. Eskinazi and J. Brennan at Syracuse University (Figures 83 and 84). There are about 75 similar small hydroelectric plants (averaging 10 megawatts) in the region, according to Eskinazi.

How much of the country's hydroelectric capacity could be used as storage for alternative energy sources is now an open question, since it has not been systematically examined. After a careful review of all requirements, Hightower found that about one-tenth of the Colorado regional capacity was available. Since his approach was quite conservative, it may not provide a bad rule of thumb, recognizing that some facilities may allow more and others less. According to that rule, there would be at least 6000 megawatts of storage capacity presently in the United States.

Hydroelectric capacity is still growing. New projects in the works in the United States will add 5000 megawatts more. In addition, Canada is one of the world's most extensive users of hydroelectricity, and one new Canadian project, at James Bay, may eventually grow to a size of 10,000 megawatts by itself. Small hydroelectric dams could be built at as many as 47,000 sites in the United States, producing 54,000 megawatts, according to a recent survey by the Army Corps of Engineers. If new solar sources were wisely coordinated with existing (and planned) hydro projects, solar energy could go quite a way before the storage potential was exhausted. But the use of existing storage is only one option.

Another option is to use solar energy systems with fossil fuel backup systems, thus saving fuel whenever the sun or wind is out. Fossil fuel backup systems are quite inexpensive compared to storage systems, so this mode of operation— usually called a fuel-saver mode—makes good sense in a time of abundant but expensive oil. Numerous analyses indicate that it is generally the first mode of solar energy deployment that will break even.

In a home, the fuel-saver mode might mean using a solar water heater with a backup of gas—a symbiosis that is favored by a number of gas companies. It might also mean using a solar heating system with an oil furnace as a backup. In an industrial application, where solar thermal systems would be used to produce steam or process heat, it would mean using an oil-fired backup system instead of storage. This symbiosis is particularly attractive because solar energy systems have high capital costs but no fuel requirements, while backup oil systems have very low capital costs and high fuel costs. In many cases, the backup oil system already exists as the present heat or steam supply, thus reducing the investment required.

In much the same fashion, solar energy units can be tied into an electric grid so that they operate in a fuel saver mode. Wind power, because it is much the cheapest source of solar electricity, is currently best suited for such a mode. The wind system would be designed to supplant power otherwise produced by the fossil-fueled (particularly oil-fired) units that the utility uses to produce electricity during periods of peak and intermediate demand. The wind system would not replace any conventional units in the fuel-saver mode, however. The utility would maintain a full complement of intermediate and peaking units to serve as backup when the winds were calm, and the benefit would derive from the fuel saved when the winds were blowing.

Restricting wind (or other solar electric sources) to peak and intermediate load periods sounds as if it might severely limit the solar energy contribution. But 40 percent of U.S. electricity is produced by intermediate and peaking units,

Figure 82. Aerial view of the Glen Canyon Dam, one of the hydroelectric facilities proposed for use in conjunction with fluctuating solar energy sources to provide storage. The view is upstream toward the Wehweap embayment. Only the excess capacity of the dam would be used to provide storage. The facility is part of the Colorado River Storage Project, a five-state project operated by the Bureau of Reclamation. A review of the excess hydroelectric capacity in other Bureau of Reclamation systems in the West indicates that it is sufficient to provide storage for 1500 to 2000 megawatts of wind (or other solar electric) power, and there is enough in the Bonneville Power Administration hydroelectric system to provide another 2000 megawatts. [Stanley Hightower and Abner Watts, Bureau of Reclamation, Denver]

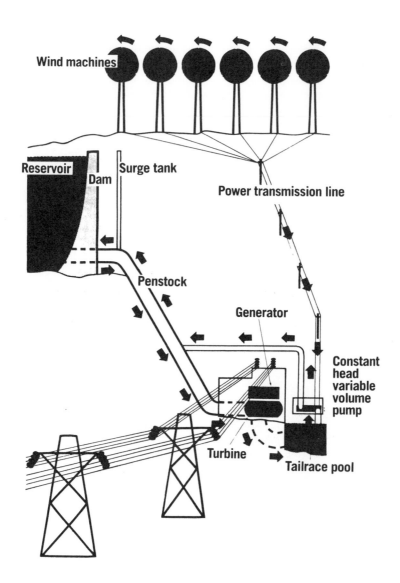

Wind machines

Reservoir

Dam

Surge tank

Power transmission line

Penstock

Generator

Constant head variable volume pump

Turbine

Tailrace pool

Figure 83. A wind-hydro system proposed for use by the Niagara Mohawk Power Corporation in upstate New York. Rather than hook up directly with the electric grid, the wind machines would be linked by a short transmission line to an electric pump at the bottom of the hydroelectric dam. Water flowing out of the reservoir through the sluice gate would produce 60-cycle power for the electric grid. The wind turbines would produce direct current to pump water uphill. The system can be used in periods of low water supply, and decoupling the wind turbine from the electric system also has the advantage that the turbine can turn at its optimum speed in any wind without the need to be synchronized with the electric power in the grid. [S. Eskinazi and J. Brennan, Syracuse University]

Figure 84. Comparison of the annual winds and the annual water supply near Polaski, New York. The wind-energy flux is highest in the October through May period, while the water supply is relatively low all year except for April and May. This is a situation common in many parts of the country where most of the annual water supply is derived from the spring runoff. [S. Eskinazi and J. Brennan, Syracuse University]

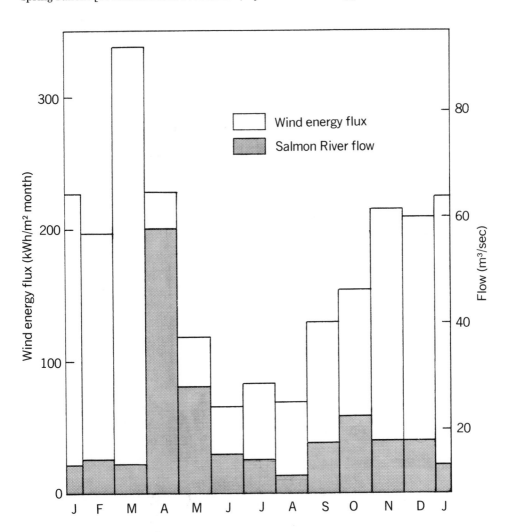

and various studies show that wind operated in a fuel-saver mode could produce 5 to 6 quads of electricity by 2020 (Appendix 1). About half the U.S. utilities have their peak loads in the winter, and the other half (generally those in the sun belt) have load curves that peak in the summer. Optimal matching of different solar sources to different utility load curves is a subject that deserves much more analysis. But it appears likely that wind power (which generally shows higher flux in the north and is invariably stronger in winter) will be most suitable for winter-peaking utilities, while solar thermal electric or photovoltaic installations (which benefit from the maximum solar insolation in summer) will be most suitable for summer-peaking utilities.

Rather than being linked through the electric grid, the fossil fuel backup for a solar system could be located at the same site as the solar electric plant and additional savings could be accrued through common usage of certain pieces of equipment. This is a feature of the particular power tower design that has been proposed by the Electric Power Research Institute (Chapter 2).

It is not clear, however, that the only benefit from a significant number of solar electric units would be fuel savings. Reliability might also be increased. Power generated at a particular site is likely to be unpredictable, but if wind or photovoltaic units were dispersed over a number of sites far enough apart that the natural energy flux was uncorrelated, the output of the connected array would be considerably more predictable. The effect of interconnection would be to average the fluctuations at different sites, with the result that the interconnected system could be counted on to reliably produce at some fraction of its potential capacity most of the time.

Again, wind is the resource for which this possibility has begun to be investigated, and preliminary indications are that—depending on the region—an interconnected wind system may earn a "capacity credit" equal to 10 to 30 percent of the total wind system capacity. This means that without any storage, a large wind system that produced 500 megawatts at peak might be able to replace 50 to 150 megawatts of conventional capacity in an electric grid.

The data base for wind correlation effects is surely modest, but the few cases that have been studied so far indicate that interconnection yields considerable improvement. J. Molly, at the Institute for Building and Construction Research in Stuttgart, studied the effects of connecting a number of windmills placed at various weather stations in West Germany and found that whereas one wind machine might be idle 35 percent of the time, the entire network would be idle less than 1 percent of the time. More importantly, Molly found that the hypothetical network, which consisted of 18 sites with an average spacing of 240 kilometers, produced more than half its average power output 65 percent of the time. A single unit met the same criterion only 35 percent of the time. This figure is of interest because 65 percent is approximately the reliability rating of a large conventional plant. Similar results were found by C. G. Justus at Georgia Tech, analyzing U.S. National Weather Service data for New England and the Midwest. Justus found that for each 1125-kilowatt wind generator in a connected system, 200 kilowatts of power was available 77 to 93 percent of the time.

The actual calculation of a capacity credit is more complicated because it depends on the characteristics of the existing utility system as well as those of the wind system. Using a system reliability model commonly employed by utilities, Ed Kahn, at Lawrence Berkeley Laboratory, studied the amount of capacity credit that could be relied on in the California networks. He found that for the installation of multi-unit wind systems sized up to several thousand megawatts, the credit would be 17 to 26 percent. But as more hypothetical wind capacity was integrated into the system, the additional capacity credit declined because there was a statistical limit to the amount of power smoothing that could be achieved. When wind power grew to constitute 25 percent of the system, Kahn found that the capacity credit for new machines dropped to zero. (The value of a wind system as a fuel saver would be governed by different considerations and would not necessarily cease at the same point.) Thus his studies indicate that the strategy of interconnecting intermittent sources through a grid to circumvent storage is particularly applicable when solar sources make up a reasonably small fraction of the energy mix.

None of this discussion should be taken as an argument against storage research. The use of intermittent energy resources is likely to grow. If the ensuing energy debate leads to a public consensus that the country should move toward a predominantly solar energy economy, energy storage will be crucial. Even in the near term, storage is essential for many remote applications and for overnight applications where the source is out of synch with the load. Furthermore, it is no doubt an oversimplification to portray the alternatives as either no storage or much storage. The growth in the use of storage is likely to be gradual but continuous, not only because of the needs of solar energy but also because of the demand and price structure of electricity.

Storage research being carried out now, largely by the Department of Energy, is building up the technical base that will be needed. Rudimentary storage is already available. Water tanks, rock beds, and specially designed pools can be used for thermal storage. Pumped hydroelectric systems and batteries can be used for electric storage. For annual storage, which means saving heat or "coolth" from one season to another, the use of larger-than-normal units in single-family houses is already being tested (Figures 85 and 86), and much larger storage systems, using lakes, reservoirs or underground aquifers, are being proposed for small communities. Many other storage ideas have been proposed, and they are in various stages of development, ranging from preliminary study to prototype testing. At least five generically different types of storage are being pursued.

Figures 85 and 86. A solar-heated house with annual energy storage in Lyngby, Denmark. Completed in 1975, the Lyngby house was designed so that it would use no energy for heating or hot water over the course of the year. Situated in a northerly latitude, the house was built with at least 1 foot of insulation, with double-glazed windows that are covered with insulated shutters at night, and with heat recovery equipment on the ventilation system. The vertical wall of solar collectors (facing page) has an area of 42 square meters. The storage tank (above) has a volume of 30 cubic meters and has 2 feet of insulation. During the first complete year's test, the house needed supplementary heating for 20 percent of its thermal load. This was due to an unusually small amount of solar radiation in the winter of the test year and greater heat leakage than expected from the house. [Torben Esbensen, Technical University of Denmark, Lyngby, Denmark]

Besides thermal storage in man-made containers, underground aquifers and lakes, bricks and ingots can also be used. Some systems using bricks are working in Europe. Rather than in sensible heat, thermal energy can be stored in the latent heat of melting in salts or in paraffin. This approach may reduce the volume of the storage device as much as 100-fold, but after several decades of research the practical problems have still not been solved.

Reversible chemical reactions can also be used to store energy. There is a growing interest in storing low-temperature energy in chemical form, but practical systems have not yet emerged. Another idea in the same category is the storage of hydrogen in metal hydrides (lanthanum, for instance). Tests of the idea are in a preliminary stage.

Mechanical and hydraulic systems would store energy by converting it from electricity into energy of compression, elevation, or rotation. Pumped storage is proven, but quite limited in its applicability by site considerations. Compressed-air storage does not appear to be under test yet in the United States (although it is being tried in Europe). The concept could be applied on a large scale using depleted natural gas fields for the storage reservoir. Alternatively, hydrogen could be stored in exhausted gas fields. Energy of rotation could be stored in flywheels, but advanced designs with high-tensile materials appear to be needed to reduce the price and volume of storage. A substantial energy penalty—up to 50 percent loss—is generally incurred by mechanical and hydraulic systems because of the inefficiency involved in a complete turnaround.

Electrochemical systems have better turnaround efficiencies but very high prices. Intensive research is now directed toward developing improved batteryies, particularly batteries that would have better weight-to-storage ratios for use in vehicles. As a successor to the lead-acid battery, sodium sulfur and lithium sulfide alternatives, among others, are being tested. A different type of electro-chemical system is the redox flow cell, so named because charging and dis-charging is achieved through reduction and oxidation reactions occurring in fluids stored in two separated tanks. To make the leading candidate—an iron redox system—competitive with today's batteries, its price would have to be at least halved.

Finally, electric energy can be stored in superconducting magnetic systems, at a high price.

Today, storage may cost anywhere from $4 per kilowatt-hour of storage capacity for thermal storage in a water tank used in conjunction with a home solar system, to $50 per kilowatt-hour for lead-acid batteries. Pumped hydroelectric storage costs approximately $10 per kilowatt-hour of capacity, considerably less than batteries. Some analysts project that thermal storage in underground aquifers could cost as little as 5 cents per kilowatt-hour of storage capacity, but the assertion is conjectural at this point. Projections for untested systems are, of course, risky, but estimates for metal hydride, hydraulic, mechanical, and super-conducting storage are in the range of $25 to $75 per kilowatt-hour. What people generally mean by a breakthrough into "cheap" storage is a cost on the order of cents per kilowatt-hour.

There are a number of promising lines of research in storage technology. Given the cost gap that needs to be spanned, it is clear that a sustained development effort is in order. But energy storage research is a conceptually rich field, and since it will be two decades or more before the long-term problems of solar energy must be faced, there are reasonable grounds for optimism. For the near-term, however, there are so many alternatives available that the lack of cheap storage should not be an impediment to the growth of solar energy.

OCEAN THERMAL ENERGY
The Biggest Gamble in Solar Power

Among the various terrestrial sytems proposed for utilizing solar energy, the largest and most complex one would use ocean water temperature differences to generate power.

Because the amount of energy that could be captured from the warm tropical oceans is truly phenomenal, the ocean thermal energy concept has many proponents. The ocean itself would be the collector and energy storage medium, eliminating the need for two of the costliest components of land-based solar thermal systems. Drawing in warm water from the surface and cold water from the depths, an ocean thermal plant could operate continuously 24 hours per day. For this reason, the ocean thermal concept has often been characterized as the only sort of solar system with the potential to provide base-load power.

Having identified a unique role for the technology, government solar energy planners put together a rapidly growing program for ocean thermal energy conversion (OTEC), which is moving quickly into expensive demonstration projects. Yet the shape and scope of the program have been determined less by solar energy administrators than by the two large aerospace companies eager to develop the concept (Lockheed and TRW), according to many observers. From $8 million 2 years ago, the budget has risen to $36 million in fiscal 1978, or about one-fifth the funding for all solar electric technologies. But electric utility companies have shown little enthusiasm for the concept and recently there has been criticism of the program's direction as well. In the words of one prominent ocean engineer, the effort appears to be going "too far too fast for what may be a tiny payoff at great price."

The OTEC concept dates back to the early part of the century, when Georges Claude tested the idea both onshore and from a ship. The shore-built plant was installed in Cuba in 1931 (Figure 87). In the late 1930s, Claude also tested the concept aboard the ship *Tunisia*, off the coast of Brazil. According to a recent review conducted by the National Academy of Sciences, the OTEC concept is technically feasible if enough money and time are spent on the project, but many critical problems remain to be solved. The review panel* found that OTEC was a long-term rather than a near-term energy option, that the projected cost estimates were "very optimistic," and that there had not been sufficient evaluation of alternative types of OTEC systems to justify the present program direction. In short, the conclusion was that the task will be harder, slower, and costlier than OTEC advocates suspect. The panel, which included a number of marine researchers and engineers, did not, however, recommend ending the program.

Similarly robust and open-minded skepticism has been displayed by the Electrical Power Research Institute, which has a $20 million, 5-year solar program but does not support any OTEC work. "We are concerned that the numbers put out seem pretty optimistic," says the coordinator of solar thermal programs, John Bigger. "We looked at the risks and decided to watch the program for now," he says. Another reason utilities are skeptical is that possible sites for

---

* The review was conducted by a panel acting under the auspices of the Marine Board of the Assembly of Engineering of the National Academy of Sciences. The members of the reviewing board were Herman Sheets (chairman), from the Department of Ocean Engineering of the University of Rhode Island, Kingston; Dillard S. Hammett, technical vice president of SEDCO, Inc., a major offshore drilling company in Dallas, Texas; John D. Issacs, a director of the Scripps Institution of Oceanography, La Jolla, California; James L. Johnston, senior economist for Standard Oil Company of Indiana, Chicago, Illinois; David G. Jopling, R&D coordinator for Florida Power & Light Company, Miami; Ralph Mitchell, from the Division of Engineering and Applied Physics, Harvard University, Cambridge, Massachusetts; Robert Panoff, a technical consultant who designed marine heat exchangers for submarines including the *Nautilus*, MPR Associates, Washington, D.C.; E. Robert Perry, who directs power transmission research for the Electric Power Research Institute, Palo Alto, California; Stanley L. Quick, manager of the steam turbine engineering department for Westinghouse Electric Corporation, Philadelphia, Pennsylvania; Warren M. Rohsenow, in the mechanical engineering department of Massachusetts Institute of Technology, Cambridge, Massachusetts; Robert E. Stein, North American director of the International Institute for Environment and Development, Washington, D.C.; and Ellis Verink, of the Department of Materials Science and Engineering department of the University of Florida, Gainesville.

OTEC are not nearly as abundant as they are for other solar technologies. Potential sites off U.S. shores are basically limited to the Gulf Coast and the lower eastern coast of Florida (although Hawaii, Puerto Rico, Singapore, Brazil, and Mexico would be ideal sites, according to a recent study by the National Science Foundation).

Having one of the few choice locations in the country, the Florida Power & Light Company has studied OTEC perhaps more than any other utility and found the arguments for it wanting. "The utilities today are faced with a high-risk world but, relatively speaking, the current concepts for OTEC are really high risk," says FPL's coordinator for research and development, David Jopling. He thinks the concept faces six to ten major technological ceilings. "Perhaps it is not impossible," he says, "but they just don't know what they are up against." Jopling, like the Academy, however, does not wish an end to OTEC.

The idea is that the program should still be analyzing the trade-offs between different OTEC concepts and building the technical base needed for more reliable cost and performance projections rather than pursuing a crash program based on one alternative. A recent evaluation by the Office of Technology Assessment warned that the present strategy is a "high risk approach" that could result in a premature choice among several concepts and result in skipping essential long-

Figure 87. An experiment by Georges Claude (left) to test the feasibility of ocean thermal energy conversion in Cuba in 1930. The most difficult part of the experiment was launching the 1750-meter cold water pipe, shown here (facing page). Two such pipes were lost to the sea before the third was deployed at a site on Matanzas Bay. The pipe was laid over the edge of a submarine cliff and reached a depth of 700 meters. Cold water from the pipe and warm water from the surface were pumped into a plant on shore, where they produced 22 kilowatts of power when the water temperatures were optimum and 12 kilowatts when leaks and seasonal current fluctuations reduced the efficiency. [Bettmann Archives and *Mechanical Engineering*]

term testing and environmental studies. The congressional assessment noted that fairly level research and development support "in the tens of millions of dollars for the next 5 to 10 years could result in a program geared toward solving important technical problems." According to Jopling, "the best promotion for the project right now is a healthy realism."

What OTEC is designed to accomplish is generally acknowledged to be the toughest problem—thermodynamically speaking—in the energy field. The temperature differences that are available between surface and deep waters in tropical latitudes are only 19° to 24°C (34° to 44°F). At such temperatures, an ideal heat engine would have an efficiency of 6 percent, and in practice OTEC will not achieve more than 2 or 3 percent efficiency. Such a meager performance rating requires that a plant draw in enormous amounts of seawater. A 100-megawatt plant would have to pump through 100,000 gallons per second—approximately as much water as flows through Boulder Dam.

It takes energy to move water around, and the pumps on an OTEC station would consume 30 percent of the gross power output of the plant, according to the engineering design studies done for the government program. Such an energy balance is astonishingly poor by the standards of present-day generating plants, which typically use less than 1 percent of their power for internal needs. Any

Figure 88. The ocean thermal energy conversion system (OTEC) proposed by Lockheed is shown in an artist's drawing. The concept is that of a spar buoy 250 feet in diameter and 1600 feet in length. Attached around the outside are four identical units that contain the pumps and turbines. These units are intended to be replaceable for repair and maintenance. Except for a helicopter landing platform, the plant would be submerged. The entire unit is designed to produce 160 megawatts of power and weighs 300,000 tons. [Lockheed Missiles and Space Co., Inc.]

slippage in the performance of the plant heat exchangers could make the OTEC energy balance even worse, and a thin layer of marine slime on the heat exchanger surfaces could push the energy balance into the red.

"Performance calculations have been based on optimal conditions which could be downgraded by any number of known and unknown effects," says the Academy review. In a recent speech given to the American Institute of Aeronautics and Astronautics, Jopling questioned whether the present OTEC designs have the ability to produce "any net energy for any appreciable length of time." The proponents of the concept are counting on the plants to operate satisfactorily with virtually no shutdowns for 20 to 40 years.

A number of different OTEC designs have been proposed. A team headed by Lockheed suggested that the plant be a submerged spar buoy (Figure 88) to make it safe against hurricanes. A team headed by TRW suggested a floating top-shaped plant (Figure 89), and a group headed by the Applied Physics Laboratory at Johns Hopkins designed a rectangular plant-ship (Figure 90) that would move slowly under its own power and have factories for electrochemicals and electro-metals, such as ammonia and aluminum, on board.

These designs have in common the fact that the plant would be built of reinforced concrete and would displace as much water as a supertanker. Suspended from the bottom of each vessel would be a pipe for taking in cold water that would be as much as 30 meters in diameter and up to 750 meters in length. The Johns Hopkins design, which is intended to develop an OTEC system as rapidly as possible by using standard components wherever possible, would use 20 pumps as large as any that are available today to draw in cold water. Twenty more pumps would draw in warm water through near-surface intakes. The Lockheed and TRW designs call for fewer pumps that would be much larger than anything yet built.

The water pumps would deliver seawater to two sets of heat exchangers. The warm-water heat exchanger would evaporate a working fluid (probably ammonia), which would produce a low-pressure gas to power a custom-designed electric turbine. To complete the thermal cycle, the ammonia would pass through a cold-water heat exchanger and be condensed back to a liquid.

Enormous heat exchangers will be required. As projected now, the surface area would cover as much as 700,000 square meters. The size depends on heat exchanger performance, which is generally recognized to be a crucial question for the OTEC program. The government program, administered by the Department of Energy, is proceeding to test a 1-megawatt heat exchanger on the old Howard Hughes–CIA special equipment barge, at a cost of $45 million (Figure 91). Performance influences cost, because the lower the performance, the bigger the heat exchangers will have to be. If the heat exchangers are built of cheap

Figure 89. Model of the OTEC design proposed by TRW. Most of the working components of the plant would float above the surface in a large circular engine room, 340 feet in diameter and 17 stories tall. Hanging below the engine room would be a 50-foot-diameter pipe to draw up cold water from below. In the TRW design (not drawn to scale), the cold-water pipe would be made of fiberglass-reinforced plastic and would be 4000 feet long. [Department of Energy]

aluminum, corrosion could be a severe problem, according to the Academy review. If they are built of expensive titanium, one OTEC plant would exhaust the entire annual national production.

Any slippage in heat exchanger performance would degrade the entire thermal cycle. This is a concern, first because the heat transfer coefficients that have been assumed in designs are superior to those attained in present seawater practice, according to Herman Sheets, who was chairman of the Academy review panel. Second, the problem of the fouling of marine hardware by slime growth has been studied extensively by industry, academia, and the military with only marginal success. Experiments conducted so far in the OTEC program indicate that a buildup of only one-fourth of a millimeter of slime would reduce the plant's performance by 60 percent. Slime grows slowly for the first 10 weeks, according to John Fetkovich at Carnegie-Mellon, but the rate may be faster in some ocean areas than others.

There appears to be no practical way to prevent slime formation, so it would have to be cleaned off—probably weekly, on the basis of available data according to Sigmund Gronich, who is head of the government program. Various types of brushes and water jets have been considered, but thorough studies have yet to be made of the cost of cleaning the huge heat exchangers.

Figure 90.   Drawing of the self-propelled OTEC plant-ship proposed by the Applied Physics Laboratory of Johns Hopkins University. The plant would not be anchored, but would "graze" in tropical waters, enabling it to operate in conditions where the temperature differential was optimum. [Applied Physics Laboratory]

Figure 91. This submersible barge, which was used in conjunction with the Howard Hughes–CIA drilling ship *Glomar Explorer* to retrieve parts of a sunken Russian submarine, has now become part of the government's OTEC research program. The 300-foot-long barge is being converted to a platform for an OTEC-type heat exchanger at a cost of about $45 million over 6 years. The heat exchanger will be designed to simulate a 1-megawatt (electric) unit. Because of the low intrinsic efficiency of OTEC plants, the thermal rating of the heat exchanger must be 40 megawatts. The contract for the heat exchanger design has been awarded to TRW, Ocean Systems and Energy Group. (But see note, p. 176.) [Department of Energy]

Other problems must be solved as well. No one has shown that it is possible to moor anything the size of an OTEC plant in the current of the Gulf Stream, and no one has tested the stresses that will be put on the huge cold-water pipe by plant motion, currents, water intake effects, and internal turbulent flow. Very little study has been made of the problem of leaks in the heat exchangers, although abundant small leaks over such a large area could be a huge problem. Finally, little consideration has been given to the reliability of the plant. Many problems, ranging from failure of the underwater transmission lines to ruptures of plumbing, or corrosive failure of key components, could shut down the plant. But the design studies fix a price of electricity or products based on the very optimistic estimate that the plant will operate at full power 90 to 95 percent of the time—a far better performance record than most land-based plants have.

Whenever shutdowns occurred, OTEC plants would face another type of fouling problem. As soon as water flow stopped for a few days, barnacles would attach to heat exchanger surfaces, according to Fetkovich, and would keep growing after operations resumed. The cost of debarnacling a whole OTEC system that accidentally became encrusted can only be guessed.

The design studies project that an OTEC plant will initially cost $1500 to $2500 per kilowatt, not counting the expense of getting electricity to shore or of building industrial plants on board. These costs are based on incomplete data, however, according to the Academy panel. Some components, such as the cold-water pipe, dwarf any marine structure that has ever been attempted. The review panel found that OTEC plants will be similar in many ways to the huge North Sea oil-drilling platforms (Condeep and Ekofisk), which have run into technical difficulties and finally cost 2 to 2.5 times what was expected. The significant point raised by many critics is that systems built for the marine environment cost many times what the same systems would cost on land. "The estimate of $2500 per kilowatt is the best that I've seen," says Jopling, "but it could be off by a factor of 5."

Many energy specialists applaud the basic studies that will provide the underpinnings of the OTEC program because these studies could lead to improved ways of using the enormous amounts of low-grade heat that are now wasted. Many power plants located on ocean coasts have sources of both hot and cold water with temperature differentials (20° to 30°F) only slightly less than the OTEC ideal. The government's uranium enrichment plants provide heat at temperature differentials equal to or greater than those at the best OTEC sites. Many critics ask why we should rush to the sea when OTEC technology could be tested more cheaply on land, and might provide cheaper power when used as a bottoming cycle on present plants. The advocates reply that no one will seriously consider investing in an OTEC plant until a true ocean-going prototype is demonstrated.

The OTEC program is scheduled to include three steps leading to a commercial-sized demonstration, now nominally set at 100 megawatts. A 5-megawatt pilot plant is due to be contracted late in 1978, followed by a 25-megawatt version and finally a commercial prototype, scheduled for installation in 1984. The program managers think that all the problems pointed out by the critics have been recognized and addressed. "We think it is an aggressive program that has set out specific review points based on the experiments necessary before proceeding to larger hardware," says Gronich. Whereas the Academy panel was under the impression that the design of the 100-megawatt system would begin in 1977, it will actually start in 1980 and "that is a big difference," according to Gronich.

Nevertheless, the breakneck pace and the technical narrowness of the program raise questions. Why is the program moving so rapidly when the burden of the technical criticism is that optimum performance from every subsystem is critical to successful operation? Why have such basic questions as how much power will be available at various distances offshore and whether and at what cost that power can be transmitted back (when undersea transmission experience is limited to rather short distances and minimal depths) not been addressed? Why such urgency to develop at great cost a solar resource which the United States has only in limited supply? Who will own and operate the plants? These questions bother not only utility executives but some solar energy advocates as well. Deploying an OTEC plant will almost certainly require more centralized institutional structures than the present array of independent utilities, which have made it clear that they will not invest in research now and are skeptical about buying the system later.

To keep on its ambitious development schedule, the OTEC program will have to grow rapidly. As noted in the congressional assessment of OTEC, the present development strategy will need "large appropriations rapidly amounting to billions of dollars." The federal investment in ocean thermal systems could easily begin to match the investments that have been made in developing nuclear power—an enterprise designed to produce a product of similar size and complexity. The gamble with OTEC is that for technical reasons alone it may provide very little energy.

Note: As this book goes to press, the Department of Energy has announced that the first test platform for OTEC, rather than the Hughes barge (Figure 91), will be the retired Navy tanker *Chepachet,* a 22,000-ton-displacement ship that will be taken out of the mothball fleet and reworked for OTEC test purposes.

APPENDIX 1
Solar Supply Projections

Table 1. Estimate by the Council on Environmental Quality of the Maximum Solar Contribution to the United States Energy Supply Assuming Accelerated Development[1] (Quads per year of displaced fuel)[2]

Technology	1977	2000[3]	2020[3]
Heating and cooling (active and passive)	small	2–4	5–10
Thermal electric	none	0–2	5–10
Intermediate temperature systems	none	2–5	5–15
Photovoltaic	small	2–8	10–30
Biomass	1.3	3–5	5–10
Wind	small	4–8	8–12
Hydropower	3	4–6	4–6
Ocean thermal energy conversion	none	1–3	5–10

[1] This estimate was based on a compilation of many sources. A full discussion is given in "Solar Energy, Progress and Promise," April 1978, available from the Council on Environmental Quality, Washington, DC 20006.

[2] A quad is a quadrillion or 10^{15} British thermal units (Btu's). One quad is equivalent to enough oil to fill 75 supertankers, or enough coal to fill 500,000 railroad cars, or enough energy to heat 500,000 homes for 20 years. In this table, electricity is converted to the equivalent fuel that would have to be burned at a power plant to supply the same amount of power. The conversion rate used is 10,000 Btu per kilowatt-hour.

[3] The estimates in these columns are not strictly additive because the various solar-electric technologies will be competing with one another.

Table 2. Projection of MITRE Corporation for the Solar Energy Division of the Department of Energy[1]: Expected Energy Savings (Quads per year of displaced fuel)[2]

Solar Technology	1985	2000	2020[3]
Hot water and space heating of buildings (direct thermal)	0.15	1.6	3–5[3]
Process heat (direct thermal)	0.02	2.0	13
Wind energy conversion systems			
Electric utility	—	1.7	6.6
Distributed	*	*	*
Solar Thermal			
Electric utility	—	0.3	2.9
Total energy systems	*	*	*
Photovoltaics			
Electric utility	—	—	0.2
Distributed	*	*	*
Ocean thermal energy conversion systems	—	—	2.4
Biomass—electricity			
Wood	—	—	0.6
Other biomass	*	*	*
Biomass—fuels			
Wood	0.03	0.4	4.4
Other biomass	*	*	*

[1] This projection was based on assumptions in the Department of Energy's National Energy Plan. It was made using MITRE's SPURR energy model. The report in which it is presented is "Solar Energy, A Comparative Analysis to the Year 2020," by G. Bennington, **ERHQ/2322–78/1**, available from the National Technical Information Service. In the table above, entries indicated by (—) are projected to supply less than 0.02 quads per year. The entries indicated by an (*) are uncertain because of the different factors that effect distributed markets.

[2] To determine the equivalent amount of primary fuel, electricity was converted to quads of fuel based on a typical efficiency of a fossil-fueled plant (10,000 Btu per kilowatt-hour).

[3] The scenario used for this energy model was that, if the entire demand were satisfied by fossil fuels, 86, 115, and 189 quads per year would be required in 1985, 2000, and 2020, respectively.

Table 3. Projection of SRI International Prepared for the Department of Energy Solar Working Group[1]: Market Penetration in Quads of Primary Energy Equivalent[2]

Solar Technology	1985	2000	2020	Technical Limit[3]
Photovoltaic	—	—	3.4	17
Thermal electric	—	—	—	10.3
Wind	—	2.0	3.8	29
Heating and cooling	0.2	4.0	6.7	19
Industrial process heat	—	—	0.1	6.3
Biomass	0.2	2.1	5.1	13
Ocean thermal energy	—	—	—	14

[1] This group was called upon to review the program balance in the solar energy division. From the projections, which were made with the SRI National Energy Model, it was concluded that wind, biomass, and heating and cooling had the best potential for the near and intermediate term, while those technologies plus photovoltaics had the best potential for the long term. The report, "Solar Energy Research and Development: Program Balance," February 1978, **HCP/M2693,** is available from the National Technical Information Service.

[2] The report developed a number of scenarios. The entries here are for the base case. For a low solar price case, most entries would only be slightly higher but wind in 2000 would double and photovoltaics in 2020 would triple. In the base case, the total consumption of primary energy in the United States would be 90 quads in 1985, 115 quads in 2000, and 140 quads in 2020.

[3] The factors governing the technical limit included the size of the resource and the size of the potential market. The photovoltaic limit was based on total intermediate- and peak-load electric demand, assuming photovoltaic would not penetrate the base-load electric market. The thermal electric limit was based on the same considerations, somewhat reduced because thermal electric systems do not pick up diffuse light. The wind limit is much larger because it includes 25 percent of the projected base-load electric market in 2020, recognizing that wind is available during the night. The heating and cooling limit is given as 80 percent of the heating load projected for all buildings in 2020, while the process heat limit is low because it includes only low temperature heat for industrial applications, and is set at 40 percent of the projected demand. Biomass is limited to 13 quads, based on an estimate of its supply. Ocean thermal energy is estimated at 14 quads, assuming it provides the total base-load demand for the South Atlantic, East South Central and West South Central regions of the United States in 2020.

Table 4. High Solar Energy Scenario Prepared by the Solar Resources Group of the Committee on Nuclear and Alternative Energy Systems (CONAES)[1] (Quads of primary energy)

Solar Technology	1985	2000	2010
Direct use			
Domestic water heating	0.2	1.2	1.3
Passive space heating	0.1	0.3	0.4
Active space heating	0.02	0.6	1.2
Nonresidential air conditioning	—	0.4	1.5
Industrial process heat	0.2	1.6	6.6
SUBTOTAL	0.5	4.1	11.0
Biomass			
Municipal Wastes	0.5	1.9	1.9
Agricultural Residues	0.5	3.5	3.5
Energy Farms[2]	(.0)	(3.0)	(3.4)
SUBTOTAL	1.0	5.4	5.4
Solar Electric[3]			
Central station	—	1.7	8.7
Total energy	—	0.5	1.9
Wind	0.1	1.4	1.8
SUBTOTAL	0.1	3.6	12.4
TOTAL	1.6	13.1	28.8

[1] The high-energy scenario assumed a vigorous policy to promote solar energy. All new buildings would be required to use solar energy after 1990 and all industrial processes would use it where applicable. The use of agricultural wastes to provide biomass would be rapidly adopted and a schedule for deploying several solar electric technologies would be set in motion. The high-energy scenario is not driven by normal economic forces. The group also prepared a low-energy scenario which assumed no policy to subsidize or otherwise hasten the implementation of solar energy. It also assumed that the costs of competing energy sources, other than natural gas, remained constant in constant dollars. The low scenario projected contributions only from biomass (0.1 quad) in 2000 and biomass and direct use (0.3 quad each) in 2010. More discussion is given in the report of the Solar Resource Group of CONAES (Washington, DC: National Academy of Sciences, 1978).

[2] The report did not include energy farms explicitly in this scenario, but developed the figures and called their inclusion a possible option.

[3] The solar electric estimate was given for central station, solar thermal total energy systems and dispersed wind systems, but the authors noted that other solar electric technologies—such as photovoltaic or ocean thermal systems—could have been substituted. The total in 2010 corresponds to 250 billion watts (GW) of central station, solar thermal plant capacity, 74 GW of total energy system electric generation, and 50 GW of wind turbines.

APPENDIX 2
Solar Collectors Available Today

Table 1. Solar Collector Manufacturers and Products

Company	Type/Feature*	Geometric concentration ratio	Wholesale price F.O.B. factory ($/m²)**	Expected price in mass production ($/m²)	Status	Design temperature
Selected stationary flat-plate collectors						
Thomason	flat plate trickle system 1 glass cover	—	32–43 (to consumer)	—	commercial	67°C
Sunworks	flat plate selective surface[a] 1 glass cover	—	85–114	about same	commercial production	57–97°C
PPG	flat plate (optional[b] selective surface) 2 glass covers Al roll-bond (optional Cu roll-bond)	—	80–107	same	commercial mass production	57–97°C
Reynolds Aluminum	flat plate (optional selective surface) 2 Tedlar covers	—	54–65	same	commercial	57–97°C
N. V. Philips	Mark II Al or Cu roll-bond covered with evacuated tubes heat mirror[c]	—	—	118	R&D	57–167°C

* Abbreviations used include m²: square meters; Al: aluminum; Cu: copper; kWe: kilowatt electric; MWth: megawatt thermal; and MWe: megawatt electric.
** Collector prices are based on inquiries in 1976, quoted in 1976 dollars. Where a price range is given, the lower price corresponds to a large order of collectors. Prices include headers, but exclude the controls, pumps and interconnecting pipes that most systems require. Prices of tracking collectors include trackers, bearings, drive motors and tracking controls.

Flat plate collector prices are quoted per square meter of gross collector area, while other collector prices are quoted per square meter of useful aperture.
[a] A selective surface is one that absorbs light well but does not radiate heat as well.
[b] Available at extra cost.
[c] A selective heat mirror coating on a transparent cover allows sunlight to enter but reflects back infrared heat trying to escape.

Table 1. (continued)

Company	Type/Feature*	Geometric concentration ratio	Wholesale price F.O.B. factory ($/m²)**	Expected price in mass production ($/m²)	Status	Design temperature
Unitspan	Cu tube and sheet 2 glass covers	—	86–92 (to consumer)	about same	commercial	57–97°C
Honeywell	flat plate Cu tube, steel sheet 2 etched AR[d] glass covers selective surface	—	145 or less	less	commercial	57–97°C
Calmac Manufacturing Corp.	flexible mat of black tubes 2 fiberglass covers	—	35–44 (retail price; some on-site fabrication required)	about same	commercial mass production	57–82°C
Stationary tubular and CPC collectors						
Owens-Illinois	evacuated glass tubes selective surface on glass inner tube white reflector	—	215 (array aperture exclusive of headers and end caps)	107–130	pilot production, demonstration and testing	97–147°C
KTA	Cu absorber glass tube half silvered	—	85–104	58–80	commercial production	57–97°C
Philips	Mark I evacuated glass tubes half silvered heat mirror coating	—	—	129	R&D	57–167°C
General Electric	evacuated glass tubes selective surface stationary external mirror	—	—	48–81	pilot production	57–167°C

[d] Antireflection coated.

Steelcraft, Inc.	Alzak mirror CPC[a] evacuated glass tube/selective surface receiver glass cover	—	269 (end of 1976; excluding rack)	161–215 (end of 1977; excluding rack)	commercial production (end of 1976)	204°C
M-7 International	solid plastic CPC for solar cells	5	—	only slightly more expensive than conventional solar cell packaging	have prototype; need capital for tooling	electricity
One-axis tracking collectors						
Albuquerque–Western, Inc.	parabolic trough Tedlar window Cu pipe	20	48–51 (including $250 tracker to drive 20–28 troughs)	(developing advanced design)	commercial production	32–100°C
Beam Engineering	parabolic trough	—	235 (or less)	"much less"	commerical production	93°C
Sandia Labs	parabolic trough evacuated glass tube/selective surface absorber	—	(see Acurex)	80% learning curve to very low price	demonstration	317°C
AAI	fixed trough tracking receiver	~8	—	54–65 (with roof credit) 86–97 (retrofit)	R&D	117°C
Hexcel, Inc.	parabolic trough Cu pipe	—	—	—	R&D	300°C
General Atomic Co.	fixed stepped trough tracking receiver	60	—	65 (installed)	R&D	497°C
Northrup, Inc.	linear Fresnel lens selective surface on Cu absorber	~10	133–180	85 (advanced design)	commercial production	93°C

[a] Compound parabolic cross-section concentrating solar collector.

Table 1. (continued)

Company	Type/Feature*	Geometric concentration ratio	Wholeslae price F.O.B. factory ($/m²)**	Expected price in mass production ($/m²)	Status	Design temperature
Sheldahl	linear heliostat	40	215–270	46–108	prototype testing	177–317°C
Itek	linear heliostat linear cavity receiver	—	—	102	R&D	537°C
Acurex Corp.	parabolic trough anodized Al mirror glass tube selective surface absorber 6 ft. wide	58	160–240	less	commercial production	60–311°C
Acurex Corp.	as above; 4 ft. wide	—	140–210	86	commercial production	60–177°C
Solartec Corp.	parabolic trough anodized Al mirror	—	100–172	about same	commercial production	204°C
Scientific Atlanta	(see General Atomic Co.) evacuated glass tube/selective surface absorber glass mirror	—	145–161	about same	commercial production	204–316°C
Two-axis tracking collectors						
ANSALDO/ Messerschmidt	ganged kinematic heliostat/tower	250–500	?	?	ready for order	600°C
E-Systems, Inc.	fixed-dish tracking receiver	—	—	50–53 installed	R&D	260°C
Sandia Labs	multiple Fresnel lens with solar cells acrylic lens	50–100	224 (without cells; with tracker; excluding design and tooling costs)	50	prototype	27–100°C and electricity

Organization	Description				Status	Temperature
JPL	parabolic dish 9.75 m² glass mirror	1000	—	115	R&D	815°C
Varian	multiple parabolic dishes with solar cells	1000	650–1000 (without cells)	very low	R&D	27–100°C and electricity
ERDA contractors	central power station heliostat/tower	—	437–492 (installed; gov't demo.)	70 (heliostats) 14 (tower receiver)	R&D	477°C
Rennssalaer Polytechnic Inst.	small ganged heliostats/tower	—	—	—	student-built project	depends on configuration
ANSALDO/Messerschmidt	parabolic dish 832 m² 100 kWe	—	1800	less	ready for order	550°C
MIT	ganged parabolic dishes with solar cells	300–500	—	?	R&D	27–100°C and electricity
ANSALDO/Messerschmidt	heliostat/tower 5 MWth/1 MWe	—	385 (including tower, boiler and tracker)	less	ready for order	600°C
Sunpower System Corp.	parabolic/trough carousel	—	129	about same	commercial production	50–260°C
Omnium-G	parabolic dish	10,000	~1000	—	commercial production	600°C

[Source: "Application of Solar Technology to Today's Energy Needs"; Office of Technology Assessment]

Table 2.   Collector Installation Costs

Collector Configuration • Components of Installation Cost	Installation Cost in $/m^2	
	Cost Components	Total
Air-cooled photovoltaics lying on roof		10.07
• installation of collectors	8.07	
• wiring	2.00	
Flat array lying on roof		18.40
• install and plumb collectors	16.40	
• wiring	2.00	
Roof replacement with air-cooled photovoltaics		1.54
• installation of collectors	8.07	
• roof credit	−8.53	
• wiring	2.00	
Roof replacement with flat array		9.87
• install and plumb collectors	16.40	
• wiring	2.00	
• roof credit	−8.53	
Air-cooled photovoltaics on frames on roof		27–47
• install collectors and frames	10–30	
• frame materials	15.39	
• wiring	2.00	
Flat array on frames on roof		27–57
• install and plumb collectors and frames	10–40	
• frame materials	15.39	
• wiring	2.00	
Tracking collectors on roof		20–40
• install and plumb collectors and frames	20–40	
Air-cooled photovoltaics on frames in field		31–51
• site preparation	.90	
• foundations	5.02	
• install collectors and frames	10–30	
• frame materials	15.39	
Flat array on frames in field		41–61
• site preparation	.90	
• foundations	5.02	
• install and plumb collectors and frames	20–40	
• frame materials	15.39	
Heliostats and air-cooled tracking photovoltaics in field		16–36
• site preparation	.90	
• foundations	5.02	
• install collectors and frames	10–30	
Plumbed trackers in field		26–46
• site preparation	.90	
• foundations	5.02	
• install and plumb collectors and frames	20–40	
Air-cooled photovoltaics raised on columns		37–57
• foundations	5.02	
• columns	6.25	
• install collectors and frames	10–30	
• frame materials	15.39	

Note: Collector installation costs vary greatly. This table is an estimate of installation costs obtainable in a mature market.

[Source: "Application of Solar Technology to Today's Energy Needs"; Office of Technology Assessment]

Table 2. (continued)

Collector Configuration • Components of Installation Cost	Installation Cost in \$/m^2	
	Cost Components	Total
Flat panels raised on columns		47–67
• foundations	5.02	
• columns	6.25	
• install and plumb collectors and frames	20–40	
• frame materials	15.39	
Heliostats and air-cooled tracking photovoltaics raised on columns		21–41
• foundations	5.02	
• columns	6.25	
• install collectors and frames	10–30	
Plumbed trackers raised on columns		31–51
• foundations	5.02	
• columns	6.25	
• install and plumb collectors and frames	20–40	

Table 3. Directory of Solar Companies

Following is the membership list of the Solar Energy Industries Association (SEIA) as of July 1978. Because the turnover of companies is high, it is important to get up-to-date information. The Department of Energy furnishes a service through a toll-free solar information number (800/523–2929) which provides names of solar manufacturers, distributers, and professional advisors on a state-by-state basis. Additional information is also available from SEIA, 1001 Connecticut Avenue, NW, Washington, DC 20036.

AAI Corp.
P.O. Box 6767
Baltimore, MD 21204

Aaron Plumbing & Mechanical Systems, Inc.
15 Poland Pl.
Staten Island, NY 10314

A Builder, Co.
900 Mark La.
Suite 305
Sheeling, IL 60090

A-1 Hydro Mechanics Corp.
94-150 Leokane St.
Waipahu-Oahu, HI 96797

Ace Irrigation & Mfg. Co.
P.O. Box 1887
Kearney, NE 68847

Acme Engineering & Mfg. Corp.
Box 978
Muskogee, OK 74401

Adv. Energy Technology, Inc.
121-C Albright Way
Los Gatos, CA 95030

Advanced Energy Systems, Inc.
P.O. Box 1916
Hagerstown, MD 21740

Aircon Ltd.
2400 Florida Ave.
Norfolk, VA 23513

Aircraftsman
P.O. Box 628
Millbrook, AL 36054

Alcan Aluminum Corp.
100 Erie View Plz.
Cleveland, OH 44110

Alien Corporation
2594 Leghorn St.
Mountain View, CA 94043

Allen Solar Center
P.O. Box 918
Ukiah, CA 95482

Allied Fabricators, Inc.
1254 Thomas Ave.
San Francisco, CA 94124

Alternate Energy Services, Inc.
R.D. 2, Box 187A
Union Bridge, MD 21791

Alternative Energy Resources
1155 Larry Mahan Dr.
El Paso, TX 79925

Aluminum Co. of America
7th St. Rd., Route 780
Alcoa Center, PA 15069

American Appliance Mfg. Corp.
2425 Michigan Ave.
Santa Monica, CA 90404

American Gas Assn.
1515 Wilson Blvd.
Arlington, VA 22209

American Heliothermal Corp.
2625 S. Santa Fe Dr.
Denver, CO 80223

American Permanent Ware
729 Third Ave.
Dallas, TX 75226

American Solar King Corp.
5808 Kingman
Waco, TX 76710

American Sunsystems, Inc.
18 High St.
West Haven, CT 06516

Ametek, Inc.
One Spring Ave.
Hatfield, PA 19440

Amtrol, Inc.
1400 Division Rd.
W. Warwick, RI 02893

Anaconda Co. Brass Div.
414 Meadow St.
Waterbury, CT 06720

Apollo Solar Systems, Inc.
5883 W. 34th St.
Houston, TX 77092

Application Engineering Corp.
850 Pratt Blvd.
Chicago, IL 60007

Appropriate Technology Corp.
22 High St., P.O. Box 975
Brattleboro, VT 05301

APS Solar Energy Corp.
4501 Curtis Ave.
Baltimore, MD 21226

Arbac Solar Engineering, Inc.
P.O. Box 4575
Downey, CA 90241

Architectural and Engineering Service Corp.
142 E. Prairie
Decatur, IL 62523

Arco Solar, Inc.
9701 Lurline Ave.
Chatsworth, CA 91311

Arizona Public Service Co.
P.O. Box 21666
Phoenix, AZ 85036

Arkla Industries, Inc.
P.O. Box 534
Evansville, IN 47704

Arneson Products, Inc.
P.O. Box 2009
Corte Madera, CA 94925

Arthur D. Little, Inc.
Acorn Park
Cambridge, MA

Artistic Construction
1127 East 3300 South, #5
Salt Lake City, UT 84106

ASG Industries
Box 929
Kingsport, TN 37662

Aton Solar Manufacturers
20 Pameron Way, #6
Novato, CA 94947

Aus-Sol Energy, Inc.
1710 S. Lamar Blvd.
Austin, TX 78704

B&B Pool Service Co., Inc.
312 Saddle River Rd.
Monsey, NY 10952

B&H Associates
12th Green St.
Allentown, PA 18102

Ballard Concrete, Inc.
Box 7175, Brandwood Station
Greenville, SC 29610

Bateman & Son, Inc.
2004 Rhode Island Ave., NE
Washington, DC 20018

Belfast Specialties Co.
P.O. Box 501, Wigwam Rd.
Belfast, NY 14711

Benz Enterprises
7105 Panorama Dr.
Rockville, MD 20855

Berry Solar Products
P.O. Box 327
Edison, NJ 08817

Bill's Electric
407 Wyman St.
Sheldon, IA 51201

Boeing Aerospace Co.
P.O. Box 3999, MS 8A-04
Seattle, WA 98124

Borg-Warner Corp.
York Division
P.O. Box 1592
York, PA 17405

Braden Steel Corp.
P.O. Box 2619
Tulsa, OK 74101

Brandem Associates
27048 Mt. Meadow Rd.
Escondido, CA 92026

Brenham Roofing & Metal Co.
P.O. Box 613
Brenham, TX 77833

Buffalow's Inc.
1245 Space Park Way
Mountain View, CA 94043

Builder Services Co.
Route 1, Box 164B
Beverly, WV 26253

Bundy Tubing
12345 E. 9 Mile Rd.
Warren, MI 48073

Burnham Corporation
P.O. Box 3079
Lancaster, PA 17604

Business of Technology, Inc.
2800 Upton St., NW
Washington, DC 20008

Butler Ventamatic Corp.
P.O. Box 728
Mineral Wells, TX 76067

California Gas Co.
P.O. Box 3249-Terminal An-
nex
Los Angeles, CA 90051

Callefaccion Y Ventilacion
Prolong Galle 18
Mexico 18, D F

Calmac Manufacturing Corp.
Box 710
150 S. Van Brunt St.
Englewood, NJ 07631

Capture Energy Products,
Inc.
1950 S. Cherokee St.
Denver, CO 80123

Carlin Co.
912 Salas Deane Hwy.
Wethersfield, CT 06109

Cavalier Energy Co.
P.O. Box 2617
Salisbury, MD 21801

Central Coast Solar Systems
1342 Gardow St.
San Luis Obispo, CA 93401

Central Valley Solar/Solahart
5057 E. McKinley
Fresno, CA 93714

Chamberlain Manufacturing
Corp.
845 Larch Ave.
Elmhurst, IL 60126

Champion Home Builders Co.
5573 North St.
Dryden, MI 48428

Chemotronics International,
Inc.
2231 Platt Rd.
Ann Arbor, MI 48104

City of Flint Department of
Community Development
1101 S. Saginaw St.
Flint, MI 48502

Climax Molybdenum Co.
3072 One Oliver Plaza
Pittsburgh, PA 15222

Cole Solar Systems, Inc.
440A East Saint Elmo Rd.
Austin, TX 78745

Columbia Technical Corp.
55 High St.
Holbrook, MA 02343

Comfort Energy Systems
1100 Blue Gum
Anaheim, CA 92806

Comfort Master of Sacra-
mento
1517 19th St.
Sacramento, CA 95814

Conenco, Inc.
Rt. 2, Yacht Cove Rd.
Columbia, SC 29210

Conserdyne Corp.
4437 San Fernando Rd.
Glendale, CA 91402

Conservation Systems, Inc.
325 W. 23rd St.
Baltimore, MD 21211

Consolidated Nat. Gas Svc.
Co.
11001 Cedar Ave.
Cleveland, OH 44106

Copper Development Assn.,
Inc.
1011 High Ridge Rd.
Stamford, CT 06905

Cosco Supply, Inc.
841 Cooke St.
Honolulu, HI 96813

Coverdale Engineering Co.
2605 S. 126th St.
Omaha, NE 68144

Creighton Solar Concepts
662 White Head Rd.
Lawrenceville, NJ 08648

Daniel Enterprises, Inc.
P.O. Box 2370
La Habra, CA 90631

Daughtry & Co.
715 W. Elm St.
P.O. Box 666
Arlington Heights, Il 60006

Daystar Corp.
90 Cambridge St.
Burlington, MA 01803

De Hart Air Conditioning, Inc.
1201 South 4th St.
Chickasha, OK 73018

Dellingers Dept. Store, Inc.
Box 758
Newton, NC 28658

Delta-T Corp.
720 Church St.
Huntsville, AL 35801

Deltair Solar Systems, Inc.
Route 2, Box 53-D
Chaska, MN 55331

Desert Sunshine Exposure
Tests
Box 185, Black Canyon Station
Phoenix, AZ 85020

Direct Heat Solar Systems
1213 Howell St.
Anaheim, CA 92805

Distinctive Home
P.O. Box 191
Greenville, TX 75401

Donohue Service Co., Inc.
4432 S. 74th East Ave.
Tulsa, OK 74145

Dow Chemical, USA
B-1605
Freeport, TX 77541

Dow Corning Corp.
Mail #CO2314
Midland, MI 48640

E. I. DuPont de Nemours &
Co.
Chestnut Run
Wilmington, DE 19898

East Greenbush Plumbing &
Htg.
34 Old Troy Rd.
E. Greenbush, NY 12061

Eclipse Solar Energy
28 Kulick Rd.
Fairfield, NJ 07006

Eco-Tek, Inc.
P.O. Box 1154
Fitzgerald, GA 31750

Ecothermia, Inc.
550 E. 12th Ave.
Denver, CO 80203

Edison Electric Institute
1140 Conn. Ave., NW
Washington, DC 20036

Edith Shedd & Associates, Inc.
Rt. 2, Box 61A1
Monroe, GA 30655

Elcam, Inc.
5330 Debbie Lane
Santa Barbara, CA 93111

Emco Building Supply, Inc.
P.O. Box 308
Bethany, OK 73008

Energex Manufacturing Corp.
4227 S. 36th Place
Phoenix, AZ 85040

Energy Advisers
P.O. Box 3473
1509 Wyldewood Dr.
Madison, WI 53704

Energy Conservation System,
Inc.
Box 1254
Fargo, ND 58102

Energy Conservation Systems
Rt. 1, Box 100
Trufant, MI 49347

Energy Converters, Inc.
2501 N. Orchard Knob Ave.
Chatanooga, TN 37406

Energy Design Associates, Inc.
P.O. Box 14187
Gainesville, FL 32601

Energy Dynamics Corp.
6062 E. 49th St.
Commerce City, CO 80020

Energy Engineering, Inc.
1901 6th St., Box 1156
Tuscaloosa, AL 35401

Energy Products
8736 Production Ave.
San Diego, CA 92126

Energy Savers, Inc.
P.O. Box 156
Arbuckle, CA 95912

Energy Savings Unlimited,
Inc.
P.O. Box 18142
Oklahoma City, OK 73118

Energy Solutions, Inc.
Hwy 93, P.O. Drawer J
Stevensville, MT 59870

Energy Systems, Inc.
4570 Alvarado Canyon Rd.
San Diego, CA 92120

Energy Systems Products,
Inc.
West Shore Plaza
12th & Market Sts.
Lemoyne, PA 17043

Enersol Company
1800 1st Int'l Bldg.
Dallas, TX 75270

Enertek Energy Engineering
P.O. Box 30
Trumann, AR 72472

Engineered Equipment Co.,
Inc.
9900 Pelumm Rd.
Lenexa, KS 66215

Environmental and Ecological
Products, Inc.
9130 Red Branch Rd.
Columbia, MD 21045

Environmental Energy
121 Broadway, Suite 535
San Diego, CA 92101

Environmental Energy Sys-
tems
2662 Pacific Park
Whittier, CA 90601

Environmental Research &
Technology, Inc.
696 Virginia Rd.
Concord, MA 01742

Eppley Laboratory, Inc., The
12 Sheffield Ave.
Newport, RI 02840

Ergonomy, Inc.
Eastview Park
7979 Victor-Pittsford Rd.
Victor, NY 14564

Everlasting Energy Systems
Box 593
Ligonier, PA 15658

EVOG
Box 36
Hebron, NH 03241

Fafco, Inc.
235 Constitution Dr.
Menlo Park, CA 94025

Fafco-South Texas
4303 Eagle Nest
San Antonio, TX 78233

Ferro Corp.
One Erieview Plaza
Cleveland, OH 44114

Flagala Corp.
9700 W. Hwy. 98
Panama City, FL 32401

Fleets Bay Enterprises
Box 994, Chesapeake Dr.
White Stone, VA 22578

Ford Products Corp.
Ford Products Rd.
Valley Cottage, NY 10989

Ford, Bacon & Davis
P.O. Box 8009
375 Chipeta Way
Salt Lake City, UT 84108

Future Systems, Inc.
12500 W. Cedar Dr.
Lakewood, CO 80215

Gasco, Inc.
P.O. Box 3379
Honolulu, HI 96842

General Electric Co.
P.O. Box 8661
Philadelphia, PA 19101

General Energy Devices
1753 Ensley Ave.
Clearwater, FL 33516

General Solar Power Corp.
99 Park Ave.
New York, NY 10016

Gleason Supply Co.
9 Water St.
Biddeford, ME 04005

Goettl Bros. Metal Products,
Inc.
2005 E. Indian School Rd.
Phoenix, AZ 85016

Gorman-Lavelle Plumbing,
Co.
3459 East 52nd Place
Cleveland, OH 44127

Graham Bros, Inc.
P.O. Box 276
Montpelier, ID 83254

Gramer Industries, Inc.
5441 E. Nassau Circle
Englewood, CO 80110

Great Lakes Solar Engr. Co.
676 N. Dearborn St.
Chicago, IL 60610

Greenfield Builders & Erec-
tors
RR 3, Box 21-A
Greenfield, IN 46140

Groundstar Energy Corp.
10 Middle St.
Bridgeport, CT 06604

Grumman Corp.
4175 Veterans Memorial Hwy.
Ronkonkoma, NY 11779

Grundfos Pumps Corp.
2555 Clovis Ave.
Clovis, CA 93612

Gulf Thermal Corp.
Amer. Alloys & Refractories
800 Oak St.
Winnetka, Il 60093

Hansberger Refrigeration &
Electric Co.
2450 8th St.
Yuma, AZ 85364

Harris Corp.
P.O. Box 1277
Kilgore, TX 75662

Harrison Radiator
A&E Bldg.
Lockport, NY 14094

Hatic Heating, Refrigeration
& Air Conditioning
P.O. Box 221
Milford, OH 45150

Hawthorne Industries, Inc.
1501 S. Dixie Hwy.
W. Palm Beach, FL 33401

Helios International Corp.
2120 Angus Rd.
Charlottesville, VA 22901

Heliosystems
3407 Ross Ave.
Dallas, TX 75204

Heliotherm, Inc.
Swedesford Rd.
Ambler, PA 19002

Hexcel Corp.
11711 Dublin Blvd.
Dublin, CA 94566

Hitachi Chemical Co. Amer.
Ltd.
437 Madison Ave.
New York, NY 10022

Hitek
42673 Ames Creek Dr.
Sweet Home, OR 97386

Honeywell, Inc.
Aerospace & Defense Group
2600 Ridgeway Pkwy.
Minneapolis, MN 55413

Honious Electric, Inc.
846 Brown St.
Dayton, OH 45409

Hydronic Institute
P.O. Box 262
35 Russo Place
Berkeley Heights, NJ 07922

Impac Sub/Decker Mfg., Co.
P.O. Box 365
Keokuk, IA 52632

Independent Energy, Inc.
P.O. Box 732
East Greenwich, RI 02818

Inter-Island Expediting
2414 Makiki Hgts. Dr.
Honolulu, HI 96822

Intersol
804 Poinciana Dr.
Gulf Breeze, FL 32561

Intertechnology/Solar Corp.
P.O. Box 340
Warrenton, VA 22186

ITT Fluid Handling Division
4711 Golf Rd.
Skokie, IL 60076

J&J Solar, Inc.
7273 N. Central
Phoenix, AZ 85020

Jamak, Inc.
1401 N. Bowie Dr.
Weatherford, TX 76086

Jason Jerusalem, Inc.
562 Sackett St.
Brooklyn, NY 11217

Jem Solar
23 Coldbrook Dr.
Cranston, RI 02920

Joe Funk Construction Engr.,
Inc.
11226 Indian Trail
Dallas, TX 75229

John E. Mitchell Co.
P. O. Box 1811
Dallas, TX 75221

Kalwall Corp.-Solar Compo-
nents
P.O. Box 237
Manchester, NH 03105

Kawneer Company, Inc.
1105 N. Front St.
Niles, MI 49120

Kendon Manufacturing, Ltd.
355 Johnson Ave. W.
Winnipeg, Manitoba
Canada

Kennard Industries, Inc.
10961 Olive Blvd.
St Louis, MO 63141

Lardic Solar Energy, Inc.
400 Agnew Rd.
Jeanette, PA 15644

Lead Industries Assoc., Inc.
292 Madison Ave.
New York, NY 10017

Leeper Heating & Cooling
107 W. 14th St.
Owensboro, KY 42301

Lennox Industries, Inc.
1600 Metro Dr.
Carrollton, TX 75006

Libbey-Owens-Ford Co.
Solar Energy Systems
1701 E. Broadway
Toledo, OH 43605

Low Income Housing Devel-
opment Corp.
3700 Chapel Hill Blvd.
Durham, NC 27707

Lundeen Corp.
2701 E. Chapman Ave.
Fullerton, CA 92631

M. C. Square
P.O. Box 6685
Fort Worth, TX 76115

Magic-Aire Division, United
Electric Co.
P.O. Box 5148
Wichita Falls, TX 76307

Mann-Russell Electric, Inc.
1401 Thorne Rd.
Tacoma, WA 98421

Marshall Industries, Inc.
576 Hazlitt Ave.
Fort Lee, NJ 07024

Martin Processing, Inc.
P.O. Box 5608
Martinsville, VA 24112

McDonnell Douglas Astro-
nautics
5301 Bolsa Ave.
Huntington Beach, CA 92637

MCM Enterprises
P.O. Box 7707
Stanford, CA 94305

Meenan Oil Company,
Inc.
113 Main St.
Tullytown, PA 19007

Miami Shores
298 N.W. 105th St.
Miami, FL 33150

Mid-Western Solar Systems,
Inc.
P.O. Box 2384
2235 Irvin Cobb Dr.
Paducah, KY 42001

Miller & Sun Heating Service
R.R. 2
Cedar Rapids, IA 52401

Mills Products, Inc.
33106 W. 8 Mile Rd.
Farmington, MI 48024

Milton Roy Co/Hartell Division
70 Industrial Dr.
Ivyland, PA 18974

Monroe Solar Energy Systems
27 Lancet Way
Brockport, NY 14420

Monsanto Co.
800 N. Lindbergh Blvd.
St. Louis, Mo 63166

Moore Fuels, Inc.
525 N. High St.
Millville, NJ 08332

Motorola, Inc.
Solar Operations
4039 E. Raymond, M-6
Phoenix, AZ 85040

Mueller Brass Co.
1925 Lapeer Ave.
Port Huron, MI 48060

Murco, Inc.
P.O. Box 1882
Monroe, LA 71201

Nat. Assn. Mutual Insurance
Co.
7931 Castleway Dr.
Indianapolis, IN 46250

National Solar Corp.
Novelty Lane
Essex, CT 06426

Natural Energy Corp.
1001 Conn. Ave., NW
Suite 530
Washington, DC 20036

Natural Power, Inc.
Francestown Turnpike
New Boston, NH 03070

Natural Resources Generation
3761 Eastern Ave., SE
Grand Rapids, MI 49508

Naturgy, Inc.
Box 3403
Tulsa, OK 74101

New England Power Co.
20 Turnpike Rd.
Westboro, MA 01581

New Jersey Aluminum Co.
P.O. Box 73
N. Brunswick, NJ 08902

North Amer. Philips Corp.
100 E. 42nd St.
New York, NY 10017

Northrup, Inc.
302 Nichols Dr.
Hutchins, TX 75141

Nova Solar Corporation
2016 Main, #2502
Houston, TX 77002

N.W. Solar Ins. of Tech.
Rt. 1, Box 114
Long Beach, WA 98631

Oden Brothers, Inc.
310 East Fourth St.
Frederick, MD 21701

Olin Brass
Olin Corporation
East Alton, IL 62024

Olympic Plating Ind., Inc.
208 15th Street, NW
Canton, OH 44707

One Design, Inc.
Mountain Falls Rt.
Winchester, VA 22601

Optical Coating Lab., Inc.
2789 Giffen Ave.
Santa Rosa, CA 95402

Owens-Corning Fiberglass
Corp.
Fiberglass Tower
Toledo, OH 43659

Owens-Illinois
P.O. Box 1035
Toledo, OH 43666

Paeco Industries
213 S. 21st St.
Birmingham, AL 35233

Parallax, Inc.
P.O. Box 180
Hinesburg, VT 05461

Parco Building Corp.
120 Parker Lane
Virginia Beach, VA 23454

Parsons Brinckerhoff, Inc.
250 W. 34th St.
New York, NY 10001

Patterson Solar Systems
185 Magnolia Ave.
Floral Park, NY 11001

Peerless Heater Co.
Spring & Schaeffer Sts.
Boyertown, PA 19512

Perficold
1529 Merrifield
Niles, MI 49120

Phelps Dodge Inc.
300 Park Ave.
New York, NY 10022

Plessey, Inc.
320 Li Expressway South
Melville, NY 11746

Poncherosa of Tamarisk
2295 Amarillo Way
Palm Springs, CA 92262

Portco
112 W. Jessamine
Ft Worth, TX 76110

Portland General Electric Co.
121 S.W. Salmon St.
Portland, OR 97204

PPG Industries
One Gateway Center
Pittsburgh, PA 15222

Precision Industries Ltd.
928 Kaamahu Place
Honolulu, HI 96817

Prof. Energy Consultants, Inc.
P.O. Box 401
Rochelle, IL 61068

Project Sun
P.O. Box 93
Mamaroneck, NY 10543

Pub. Serv. Co. of New
  Mexico
P.O. Box 2267
Albuquerque, NM 87103

R.M. Thornton, Inc.
9200 Edgeworth Dr.
Capital Heights, MD 20027

R.M. Warren Co.
6017 Staples Mill Rd.
Richmond, VA 23228

Radco Prod., Inc.
2877 Industrial Pkwy.
Santa Maria, CA 93454

Raleigh Solar Systems
8454 N.W. 58th St.
Miami, FL 33166

Rane Air Solar Enterprises,
  Inc.
16417 Rinaldi St.
Granada Hills, CA 91344

Raypak, Inc.
31111 Agoura Rd.
Westlake Village, CA 91361

Raytheon Company
141 Spring St.
Lexington, MA 02173

Realtec, Inc.
P.O. Box 80
Sapphire, NC 28774

Research Products Corp.
P.O. Box 1467
Madison, WI 53701

Resource Alternatives
P.O. Box 175
Mossville, IL 61552

Revere Copper & Brass, Inc.
2107 S. Garfield Ave.
P.O. Box 3246, Terminal Ann.
Los Angeles, CA 90040

Revere Solar & Arch Products
P.O. Box 11168
1200-26 E. 40th St.
Chattanooga, TN 37401

Rex Plumbing & Heating
8 End St.
Kingston, PA 18704

Reynolds Metals Co.
6601 W. Broad St.
Richmond, VA 23261

Rheem Water Heater Div.
7600 S. Kedzie Ave.
Chicago, IL 60652

RHO Sigma, Inc.
11922 Valerio St.
N. Hollywood, CA 91605

Rhodia, Inc.
1145 Towbin Ave.
Lakewood, NJ 08701

Richdel, Inc.
1851 Oregon St.
P.O. Drawer A
Carson City, NE 89701

Robertshaw Controls Co.
Long Beach Blvd. at Long
  Beach Freeway
Long Beach, CA 90802

Rohm & Haas Co.
Independence Mall West
Phialdelphia, PA 19105

S-Systems, Inc.
515 Lookout, #211
Richardson, TX 75080

San Diego Gas & Electric
P.O. Box 1831
San Diego, CA 92112

Sanders Associates, Inc.
D. W. Highway South
Mail Stop NHQ 1-573
Nashua, NH 03061

Scientific Atlanta, Inc.
3845 Pleasantdale Rd.
Atlanta, Ga 30340

Sealed Air Corp.
2015 Saybrook Ave.
Commerce, CA 90040

Self-Sufficient Living
  Center
P.O. Box 4387
Tulsa, OK 74104

Sennergetics
18621 Parthenia St.
Northridge, CA 91324

Sensor Technology, Inc.
21012 Lassen St.
Chatsworth, CA 91311

SES, Inc.
Tralee Industrial Park
Newark, DE 19711

Sidney Scarborough Co.
Rt. 1, Box 720
Biloxi, MS 39532

Sierra Solar Systems, Inc.
P.O. Box 310
Nevada City, CA 95959

Signa Corporation
P.O. Box 501
Decatur, IN 46733

Silgas, Inc.
Box 546
Jeffersonville, IN 47130

Silicon Material, Inc.
341 Moffett Blvd.
Mountain View, CA 94043

Siltec Corporation
3717 Haven Ave.
Menlo Park, CA 94025

Singer Co.
Room 6013C
30 Rockefeller Plaza
New York, NY 10020

Skidmore, Owings & Merrill
30 W. Monroe St.
Chicago, IL 60603

Slack Associates
540 S. Longwood St.
Baltimore, MD 21223

Smacna
Tysons Corner
8224 Old Courthouse Rd.
Vienna, VA 22180

Sol Energy Corp.
761 Rhode
Hillside, IL 60162

Sol. Energy, Inc.
4254 Cleveland Ave.
Ft. Myers, FL 33901

Sol Power Engineering Co.
9821 Westland Dr.
Concord, TN 37720

Solar Age Systems
217 East Hickory
P.O. Box 1983
Denton, TX 76201

Solar Applications, Inc.
4977 Chaparral Way
San Diego, CA 92115

Solar Building Systems
610 W. Broadway, Suite 209
Tempe, AZ 85282

Solar Collectors of Santa Cruz
2902 Glen Canyon Rd.
Santa Cruz, CA 95060

Solar Constructioneers
21 Fundus Rd.
W. Orange, NJ 07050

Solar Contractors of Okla-
   homa
Kent Construction Co.
P.O. Box 640
Henryetta, OK 74437

Solar Control Corporation
5595 Arapahoe Rd.
Boulder, CO 80302

Solar Crafters, Inc.
Tsali Lane, Rt. 2
Strawberry Plains, TN 37871

Solar Development, Inc.
Garden Industrial Park
3630 Reese Ave.
Riviera Beach, FL 33404

Solar Dynamics of Arizona
Box 647
Lake Havasu City, AZ 86403

Solar Earth Energy, Inc.
2020 Brice Rd., #232
Reynoldsburg, OH 43068

Solar Electric of Oklahoma
3118 S. Sheridan Rd.
Tulsa, OK 74135

Solar Energy & Insulation Co.
P.O. Box 2084, Ellicott Sta.
Buffalo, NY 14205

Solar Energy Analysis Lab.
4325 Donald Ave.
San Diego, CA 92117

Solar Energy Associates
P.O. Box 682
Fredericksburg, VA 22401

Solar Energy Barn, Inc.
Box 107B, Rt. 9
Rhinebeck, NY 12572

Solar Energy Contractors, Inc.
P.O. 16425
3145 Leon Rd.
Jacksonville, FL 32261

Solar Energy Info. Service
18-2nd Ave., Box 204
San Mateo, CA 94401

Solar Energy News Div. of
   Pro-Growth, Inc.
178 Miller Ave.
Providence, RI 02905

Solar Energy of the Rockies
P.O. Box 29171
Denver, CO 80229

Solar Energy Products, Inc.
1208 N.W. 8th Ave.
Gainesville, FL 32601

Solar Energy Research Corp.
701 B, So. Main St.
Longmont, CO 80501

Solar Energy Systems, Inc.
Cherry Hill Industrial Park
#1 Olney Ave.
Cherry Hill, NJ 08003

Solar Energy Systems of Ga.
P.O. Box 969
Roswell, GA 30077

Solar Energytics, Inc.
P.O. Box 532
Jasper, IN 47546

Solar Environmental Engr.
   Co.
P.O. Box 1914
Fort Collins, CO 80521

Solar Etc.
3141 Laguna Shores
Corpus Christi, TX 78418

Solar Heating & Cooling
20 Community Place
Morristown, NJ 07960

Solar Heating & Cooling, Inc.
Bankers Trust Office Park
6 Pope Ave.
Hilton Head Isle, SC 29928

Solar Homes of Montana
1911 11th Ave., So.
Great Falls, MT 59405

Solar, Inc.
P.O. Box 246
Mead, NE 68041

Solar Industries, Inc.
P.O. Box 303
Plymouth, CT 06782

Solar Innovations
412 Longfellow Blvd.
Lakeland, FL 33801

Solar Investors Assoc.
325 Prospect Ave.
Mamaroneck, NY 10543

Solar Kinetics, Inc.
P.O. Box 10764
Dallas,TX 75207

Solar King
5031 E. 29th
Tucson, AZ 85712

Solar Master
8624 De Soto Ave.
Canoga Park, CA 91304

Solar Northwest Corp.
Rt. 1, Box 114
Long Beach, WA 98631

Solar of New Jersey
1 Graham Ave.
Metuchen, NJ 08840

Solar Operations
4039 E. Raymond, M-6
Phoenix, AZ 85040

Solar Power Corp.
P.O. Box 8290
New Orleans, LA 70182

Solar Products
115 Millet St. North
Dix Hills, NY 11746

Solar Products, Inc.
12 Hylestead St.
Providence, RI 02905

Solar Products Infor. & Engr.
P.O. Box 506
Columbus, NC 28722

Solar Research
Div. of Refrigeration Research
525 N. Fifth St.
Brighton, MI 48116

Solar Research, Inc.
4906 Balcones Dr.
Austin, TX 78731

Solar Research Systems
3001 Red Hill Ave., I-105
Costa Mesa, CA 92626

Solar Sales
906 Kings Mill Rd.
Chapel Hill, NC 27514

Solar Screens, Inc.
817 E. Euclid
San Antonio, TX 78212

Solar Service System, Inc.
1685 S. Chase Ct.
Lakewood, CO 80226

Solar Solutions, Inc.
Mill Village, Route 20
Sudbury, MA 01776

Solar Southeast, Inc.
P.O. Box 3826
Kingsport, TN 37663

Solar State Systems
2821 Ladybird Lane
Dallas, TX 75220

Solar Store, Inc.
2900 Monterey, SE
Albuquerque, NM 87106

Solar Supply Unltd.
4680 Shoulders Hill Rd.
Suffolk, Va 23435

Solar Systems
901 Main St.
Louisville, CO 80027

Solar Systems Devel. Corp.
P.O. Box 4369
Albuquerque, NM 87106

Solar Systems, Inc.
507 W. Elm St.
Tyler, TX 75701

Solar Systems of Va., Inc.
10754 Jefferson Ave.
Newport News, VA 23601

Solar Systems West
421 Court St.
Elko, NV 89801

Solar Technic Intl. Inc.
9459 Timberleaf Dr.
Dallas, TX 75243

Solar Technology Company
210 New Hackensack Rd.
Poughkeepsie, NY 12603

Solar Technology, Inc.
3927 Oakcliff Ind. Ct.
Atlanta, GA 30340

Solar Usage Now, Inc.
Rt 18, Box 306
Bascom, OH 44809

Solar Warehouse, Inc.
140 Shrewsbury Ave.
P.O. Box 639
Red Bank, NJ 07701

Solar West Construction Co.
565 Mountain Green Dr.
Calabasas, CA 91302

Solar West, Inc.
P.O. Box 892
Fresno, CA 93714

Solar Work
3429 Chamoune
San Diego, CA 92105

Solar World, Inc.
4449 N. 12th St., Suite 7
Phoenix, AZ 85014

Solar/Wind Energy Systems
10833 Farmington Rd.
Livonia, MI 48150

Solarcon, Inc.
607 Church
Ann Arbor, MI 48104

Solarexx
P.O. Box 575
Onancock, VA 23417

Solarflame Systems, Inc.
P.O. Box 99
Leroy, IL 61752

Solaron Corp.
720 S. Colorado Blvd.
Denver, CO 80222

Solarphonics
P.O. Box 1571
San Luis Obispo, CA 93406

Solartech Systems Corp.
P.O. Box 1650
Lubbock, TX 79408

Solarthermics Co.
Old Boston Rd.
Wilton, CT 06897

Solartran Corp.
P.O. Box 496
Escanaba, MI 49829

Solarway
P.O. Box 249
Healdsburg, CA 95448

Solcan Ltd.
RR # 3—London
Ontario
Canada N6A 4B7

Solenco Corporation
175 River Rd.
Flanders, NJ 07836

Solenergy Corp.
23 North Ave.
Wakefield, MA 01880

Solray Energy Systems, Inc.
P.O. Box 2124
Clifton, NJ 07015

Solstice Construction, Inc.
P.O. Box 7441
Ann Arbor, MI 48107

Soltek Associates
P.O. 747
Berkeley, CA 94701

Soltek Construction Co.
76 Sandy Neck Road
E. Sandwich, MA 02537

Son Solar Systems
14007 Briardale Ln.
Tampa, FL 33618

Sooner Solar
5820 E. Reno
Midwest City, OK 73110

Southeastern Solar Systems, Inc.
2812 New Spring Rd., #150
Atlanta, GA 30339

Southeastern Solar Systems of America, Inc.
P.O. Box 5567 McFarland Dr.
Tuscaloosa, AL 35401

Southern California Edison Company
P.O. Box 800
Rosemead, CA 91770

Southern California Gas Co.
Terminal Annex
P.O. Box 3249
Los Angeles, CA 90051

Southwest Solar Systems, Inc.
5555 N. Lamar, Ste. J-107
Austin, TX 78751

Speed Scientific Sch.
EE Dept.
Univ. of Louisville
Louisville, KY 40208

Startrack Solar Systems, Inc.
1835 Third Ave., SE
Cedar Rapids, IA 52403

State Industries, Inc.
By-Pass Road
Ashland City, TN 37015

Suburban Propane Gas Corp.
Box 206, Rt. 10
Whippany, NJ 07981

Sun & Roof Engineering, Inc.
11622 W. Expo Blvd., Ste. 2
W. Los Angeles, CA 90064

Sun King
2722 W. Davie Blvd.
Ft. Lauderdale, FL 33312

Sun Life Solar Products
12900 S.E. 32nd St.
Milwaukie, OR 97222

Sun Power Systems Corp.
510 S. 52nd St., #101
Tempe, AZ 85282

Sun Press International
P.O. Box 488
Niceville, FL 32578

Sun Stone Solar Energy Equip.
P.O. Box 941
Sheboygan, WI 53081

Sun Systems, Inc.
P.O. Box 155
Eureka, IL 61530

Sun Works Construction, Inc.
808 Valley Hwy.
Acme, WA 98220

Sun-Design Evolution
1903 Tulip Terrace
Manhattan, KS 66502

Sun-Pac, Inc.
P.O. Box 8169
Alexandria, LA 71301

Sun-Ra Solar Consulting
303 N.E. 56th
Seattle, WA 98105

Sun-Ray Solar Equip. Co., Inc.
415 Howe Ave.
Shelton, CT 06484

Sun-Up Solar Systems
P.O. Box 413
Charleston, MO 63834

Sunbelt Solar Systems, Inc.
2418 Tangley
Houston, TX 77005

Sunburst Energy Systems
20 W. Louise Ave.
Salt Lake City, UT 84115

Suncoast Solar, Inc.
973 Virginia Ave.
Palm Harbor, FL 33563

Sunearth Solar Products Corp.
RD #1, Box 337
Green Lane, PA 18054

Sunny Days, Inc.
P.O. Box 58565
Houston, TX 77058

Sunray Works Co.
350 Parsippany Rd., Ste. 64
Parsippany, NJ 07054

Sunrise Solar Systems, Inc.
10 Candlelight Dr.
Montvale, NJ 07645

Sunsav, Inc.
890 East St.
Tewksbury, MA 01876

Sunsaver
3034 N. 33rd Dr.
Phoenix, AZ 85017

Sunsaver Corporation
P.O. Box 276
N. Liberty, IA 52317

Sunshine Co. Inc.
501 Mannakee St.
Rockville, MD 20850

Sunshine Unltd.
900 N. Jay St.
Chandler, AZ 85224

Sunshine Ventures
6601 Merwin Rd.
Worthington, OH 43085

Sunsport Solar Products, Inc.
146 E. Main St.
Carrboro, NC 27510

Sunstruction Unlimited, Inc.
R.D. 2, Box 113
Allentown, PA 18103

Suntec Systems, Inc.
21405 Hamburg Ave.
Lakeville, MN 55044

Sunwall, Inc.
P.O. Box 9723
Pittsburgh, PA 15229

Sunworks, Inc.
P.O. Box 1004
New Haven, CT 06508

Sven Tjernagel Solar Systems
477 Woodcrest Dr.
Mechanicsburg, PA 17055

Taco, Inc
1160 Cranston St.
Cranston, RI 02920

Termosolar Comercio E Indus. Lt.
Rua Senador Alencar, 129-3Andar
Rio De Janiero RJ, Brazil

Terra Light, Inc.
128 Spring St.
Lexington, MA 02173

Texas Testing Laboratory, Inc.
P.O. Box 2144
Dallas, TX 75221

Thermal Dynamics, Inc.
2285 Emerald Court
Reston, VA 22091

Thermill Corporation
3501 N.W. 60 St.
Miami, FL 33142

Thermo Solar, Inc.
4141 Sherbrooke St. W.
Montreal, Quebec
Canada

Thermo Solar Water Heaters Co.
P.O. Box 2858
Amman, Jordan

Toltec, Inc.
40th East Main
Clear Lake, IA 50428

Tomorrow Today, Inc.
6115 28th S.E., Ste. 121
Grand Rapids, MI 49506

Tulare County Air Cond. Svc.
P.O. Box 3245
Visalia, CA 93277

Turbonics, Inc.
11200 Madison Ave.
Cleveland, OH 44102

U.S. Solar Systems
3510 N. Hwy. 97
Bend, OR 97701

Universal 100 Solar Energy Sys.
501 Elwell St.
Orlando, FL 32803

Uniwave, Inc.
75 Marine St.
Farmingdale, NY 11735

Valley Oil Co.
P.O. Box 12249
Salem, OR 97309

Vanguard Energy Sys.
9133 Chesapeake Dr.
San Diego, CA 92123

Vaughn Corp.
386 Elm St.
Salisbury, MA 01950

Vermont Solar Group
Box 292
Warren, VT 05674

Virginia Chemicals, Inc.
4100 Platinum Way
Dallas, TX 75232

Virginia Solar Comp., Inc.
Rt. 3, Hwy. 29S
Rustburg, VA 24588

Vulcan Solar Industries
P.O. Box 85
Central Falls, RI 02860

W.I.S.E. Managers, Inc.
P.O. Box 38
Sandy Creek, NY 13145

Washington Gas Light Co.
6801 Industrial Rd.
Springfield, VA 22151

Watson Pool Co., Inc.
P.O. Box 177
Easley, SC 29640

Wayne Manufacturing Corp.
VP — Engineering
P.O. Box 550
Waynesboro, VA 22980

Weather-Made Systems, Inc.
Rt. 7, Box 300D
Springfield, MO 65802

Went-Sol, Inc.
562 Main St.
East Greenville, PA 18041

Western Solar Dev., Inc.
1236 Callen St.
Vacaville, CA 95688

Western Sun Cool
6520 E. Exeter Blvd.
Scottsdale, AZ 85251

Whelchel Solar Enterprises
2050 Carroll Ave.
Chamblee, GA 30341

Wilcox Mfg. Corp. of Texas
327 County Fair Dr.
P.O. Box 38572
Houston, TX 77060

Wilmington Coil Products
602 Sunnyvale Dr.
Wilmington, NC 28401

Winsun Corp.
8030 S. Gary
Tulsa, OK 74136

Wojcik Industries, Inc.
301 N. Brandon
Fallbrook, CA 92028

Wolverine — Div. of UOP
P.O. Box 2202
Decatur, AL 35601

Wormser Scientific Corp.
88 Foxwood Rd.
Stamford, CT 06903

WW Solar Supply Co.
P.O. Box 403
Hainesport, NJ 08036

Wyle Laboratories
7800 Governors Dr. W.
Huntsville, AL 35807

Yellow Windmill Enterprises
Rt. 1, Box 40
Pallacios, TX 77465

Ying Manufacturing Corp.
1957 W. 144th St.
Gardenia, CA 90249

Zien Plumbing and Heating Co.
4450 N. Oakland Ave.
Milwaukee, WI 53211

Other Manufacturers

Acurex Corporation
485 Clyde Avenue
Mountain View, CA 94042

Albuquerque Western Industries, Inc.
612 Commanche, N.E.
Albuquerque, NM 87107

Argonne National Labs.
9700 South Cass Avenue
Argonne, IL 60439

Battelle Memorial Labs.
505 King Avenue
Columbus, OH 43201

E-Systems, Inc.
P.O. Box 226118
Dallas, TX 75266

Omnium-G
815 Orange Thorpe Park
Anaheim, CA 92801

APPENDIX 3
Commercial Wind Machines

Table 1. Small Commercial Wind Turbine Generators

Name	Size (kW)	Blade diameter (ft.)	Number of blades[1]	Rated speed (mph)	Cut-in speed (mph)	Cut-out speed (mph)	Governor	Generator type
Aero-Power (USA)	1.000	6.00	2	32.0	10.0	—	full feathering 12-volt	DC generator
Aerowatt (France) 24FP7	0.028	3.30	2	15.7	6.7	—	centrifugal pitch control	3-phase alternator
150FRP7	0.130	6.70	2	15.7	6.7	—	"	"
300FP7	0.350	10.70	2	15.7	6.7	—	"	"
1100FP7	1.125	16.70	2	15.7	6.7	—	"	"
4100FP7	4.100	30.7	2	16.0	3.3	55	"	"
American Wind Turbine SST (USA)	0.450	8.00	24	20.0	10.0	30	vane deflect	alternator or water pumper
	0.900	12.00	36	20.0	10.0	adjustable	"	"
	1.800	16.00	48	20.0	10.0	—	"	"
Amerenalt (USA)	1.500	8.00	24	30.0	9.0	adjustable	vane deflect	alternator
	2.500	8.00	24	40.0	10.0	adjustable	"	"
Domenico Sperandio & Ager (Italy)	0.250							
	0.500							
	1.000							
Dominion Aluminum DAF (Canada)	2.000	15.00	2–Darrieus	23.0	7.0	65	spoilers or induction generation	alternator
	4.000	15.00	"	23.0	7.0	65	"	"
	6.000	20.00	"	23.0	7.0	65	"	"
	8.000	30.00	"	23.0	7.0	65	"	"
Dunlite Electric Co.[2] (Australia)								
L	1.000	12.00	3	25.0	10.0	—	80° feather	alternator
M	2.000	12.00	3	25.0	10.0	none	"	"
Elektro[3] (Switzerland)								
W50	0.050	1.42	Savonius	39.0	7.0	none	—	alternator
W250	0.250	2.20	"	40.0	7.0	"	—	"
WV05	0.600	8.33	2	20.0	7.0	50	full feathering	"
WV15G	1.200	9.83	2	23.0	7.0	50	"	"

Model							Overspeed control	Generator
WV25G	1.800	11.50	2	22.0	7.0	50	"	"
WV25/3G	2.500	12.50	3	23.0	7.0	50	"	"
WV35G	4.000	14.42	3	24.0	7.0	50	"	"
WVG50G	6.000	16.42	3	26.0	7.0	50	"	3-phase alternator
Enag (France)	0.400	n/a	2	n/a	n/a	n/a	n/a	n/a
	1.200	n/a	3	n/a	n/a	n/a	n/a	n/a
	2.500	n/a	3	n/a	n/a	n/a	n/a	n/a
Jacobs (USA)								
J47	3.000	14.00	3	20.0	8.0	—	centrifugal feathering flyball	120-volt DC
J45	2.500	14.00	3	20.0	8.0	—		120-volt DC
J46	2.800	14.00	3	20.0	8.0	—	"	32-volt DC
J49	2.500	14.00	3	20.0	8.0	—	"	"
J50	2.000	14.00	3	20.0	8.0	—	"	"
J51	1.800	14.00	3	20.0	8.0	—	"	"
Kedco–1200 (USA)	1.200	12.00	3	21.0	7.0	—	centrifugal feathering	alternator
Lubing (French)	—	7.22	3	11.2	—	—		water pumper
M022–3	0.400	7.22	6	18.0	6.7	35	3 flared, 3 feathered	alternator
NOAH (Switzerland)	55.000	36.00	10 double rotor	20.0	6.7	—	electronic	28 perm. magnet AC generator
Sencenbaugh (USA)	0.750	12.00	3	20.0	8.0	30	vane deflect	14-volt DC
Winco Wind Charger 1222H (USA)	0.200	6.00	2	23.0	7.0	—	air brake	double carbon brush
Windstream–25 Grumman (USA)	15.000	25.00	3	26.0	9.0	60	n/a	alternator
Zephyr (USA)	15.000	20.00	3		8.0	45	spoilers and servo control	alternator

1 Unless otherwise noted, the machine is a horizontal-axis type.
2 Dunlite machines have been used in Australia since the 1930s.
3 Elektro also makes a 10kW machine.

[Source: Department of Energy Small-Scale Wind Program, Rocky Flats Test Center, Golden, Colorado]

Table 2. Wind Turbine Generator Distributors and Manufacturers

Aero–Power
432 Natoma Street
San Francisco, CA 94103
SMALL WTGs

American Energy Alternatives
P.O. Box 905
Boulder, CO 80302
AMERENALT

American Wind Turbine
1016 East Airport Road
Stillwater, OK 74074
AMERICAN WIND TURBINE

Automatic Power, Inc.
Pennwalt Corporation
213 Hutcheson Street
Houston, TX 77023
AEROWATT

Dyna Technology, Inc.
P.O. Box 3263
Sioux City, IA 51102
WINCO

Edmund Scientific Company
380 Edscorp Bldg.
Barrington, NJ 08007
SMALL WTGs

Energy Alternatives, Inc.
P.O. Box 233
Leverett, MA 01054
DUNLITE, WINCO, LUBING

Enertech, Inc.
P.O. Box 420
Norwich, VT 05055
DUNLITE, ELEKTRO

Environmental Energies, Inc.
P.O. Box 73, Front Street
Copemish, MI 49625
DUNLITE, ELEKTRO, WINCO,
JACOBS

Grumman Aerospace Corporation
Energy Programs
4175 Veterans Memorial
Highway
Ronkonkoma, NY 11779
WINDSTREAM–25

Independent Energy Systems
6043 Sterrettania Road
Fairview, PA 16415
DUNLITE, JACOBS

Independent Power Developer
P.O. Box 1467
Noxon, MT 59853
DUNLITE, ELEKTRO, WINCO

Kedco Incorporated
9016 Aviation Blvd.
Englewood, CA 90301
KEDCO

Northwind Power Company
P.O. Box 315
Warren, VT 05674
JACOBS, AERO-POWER, WINCO

Pinson Energy Corporation
P.O. Box 7
Marstons Mill, MA 02648
PINSON

Real Gas and Electric Company
P.O. Box "A"
Guerneville, CA 95446
DUNLITE, ELEKTRO, WINCO,
JACOBS

REDE Corporation
P.O. Box 212
Providence, RI 02901
DAF

Sencenbaugh Wind Electric
P.O. Box 11174
Palo Alto, CA 94306
SENCENBAUGH

WTG Energy Systems
Box 87, 1 LaSalle Street
Angola, NY 14006
200kW MACHINES

Windlite Alaska
P.O. Box 43
Anchorage, Alaska (AK)
99510
DUNLITE, ELEKTRO, WINCO

Zephyr Wind Dynamo Company
P.O. Box 241
Brunswick, ME 04011
ZEPHYR

APPENDIX 4
The Federal Solar Research Budget

Table 1. Department of Energy Request to Congress for 1979*—Summary of Solar Research Categories

Activity	Requested Authorization for FY 1979 (millions of dollars)
Operating expenses	
Thermal applications	
Heating and cooling of buildings	33.1
Agricultural and industrial process heat	11.0
Technology support and utilization	14.1
Solar electric applications	
Solar thermal	69.0
Photovoltaics	75.8
Wind energy	39.3
Ocean thermal	32.5
SUBTOTAL	274.8
Capital acquisition	
Not related to construction	
Thermal applications	1.0
Solar electric applications	5.4
Construction projects	
Solar electric applications	28.0
SUBTOTAL	34.4
TOTAL for Solar	309.2

* Four months after initial request, the Administration sent the Congress an amendment to the budget that reprogrammed $130 million of support that had been largely designated for nuclear energy and directed $70 million of that amount to solar energy. The announcement was made on "Sun Day," May 3, 1978. The amendment specified $30 million for photovoltaic research, $20 million for wind machine demonstrations, $5 million for passive heating and cooling, $5 million for dispersed energy system demonstrations, $5 million for appropriate technology grants, and $5 million for solar training and education. The amendment also specified $20 million for a low-head hydropower program.

Table 2. Breakdown of Expenditures for Solar Programs*—Initial Department of Energy Request to Congress

Activity	Requested Authorization for FY 1979 (millions of dollars)
Agricultural and industrial process heat	
• agricultural	3.0
• industrial	8.0
Technology support and utilization	
• environmental and resource assessment	6.5
• satellite power system concept evaluation	4.6
• technology transfer	3.0
Solar thermal power systems	
• large-scale systems and applications	16.9
• central receivers subsystems and components	10.1
• total energy systems	14.0
• small power systems (irrigation)	14.0
• advanced studies (new high temperature receivers)	14.0
Photovoltaic energy conversion	
• advanced R&D (thin films)	13.5
• systems analysis and engineering	3.9
• silicon technology development	27.0
• concentrator technology development	5.8
• mission analysis and standards	3.6
• market-pull test (effect of federal buying)	22.0
Wind energy conversion systems	
• program development and technology (mission analyses, legal/environmental studies and advanced concepts)	10.1
• farm and rural systems (the small-scale program)	9.8
• kW-scale systems (the 200-kW turbines)	1.9
• MW-scale systems	15.1
• large-scale multi-unit systems (studies)	2.4
Ocean thermal energy conversion	
• project support (conceptual engineering)	3.5
• definition planning	0.8
• engineering development	8.2
• engineering testing (using Hughes barge)	13.5
• advanced research and technology	6.5
Solar electric capital equipment	
• solar thermal	3.0
• photovoltaic	0.3
• wind energy	1.4
• ocean thermal	0.7
Construction	
• ten-megawatt central receiver pilot plant at Barstow, California	28.0

* Projected total costs for certain projects funded through operating expenses are included in the FY 1979 budget. They are $15.2 million for the total energy pilot program at Fort Hood, Killeen, Texas; $6.3 million for the photovoltaic total energy system at Mississippi County Community College in Blytheville, Arkansas; $13.5 million for the total energy process heat plant in Shenandoah, Georgia; $45.5 million for converting the Hughes mining barge to a test bed for ocean thermal energy heat exchangers; $50 million for an unspecified "small power system" that will generate 1 megawatt of electricity; $16.8 million for design and construction of the 300-foot-diameter Mod 2 wind turbine; $14.1 million for the 200-foot-diameter Mod 3 wind turbine; and $3.6 million for a 100- to 200-kilowatt Mod 4 wind turbine.

Bibliography

Chapter 1

1. *Application of Solar Technology to Today's Energy Needs* (Washington, DC: Office of Technology Assessment, 1978). The most thorough review of solar technology to date, this volume is an invaluable reference. It covers direct solar technologies, but not wind, biomass, or ocean thermal energy conversion. It may be ordered from the Superintendent of Documents, U.S. Government Printing Office, Washington, DC 20402.
2. *Bulletin of the Atomic Scientists.* A Series of 12 occasional articles on solar energy, edited by Robert H. Williams, that ran from November 1975 to October 1977. This excellent series is due to be published in book form as *Toward a Solar Civilization* (Boston: MIT Press, 1978).
3. "Introductory Annotated Solar Energy Bibliography." A good, short bibliography giving sources of solar policy discussion as well as how-to-do-it books and actual plans. Available from the National Center for Appropriate Technology, Butte, MT 59701.
4. *Solar Age* (Church Hill, Harrisville, NH 03450; $20 per year). A monthly magazine of solar energy for a broad audience.

Chapter 2

1. "Solar Thermal Conversion Mission Analysis" (El Segundo, CA: Aerospace Corporation, 1974). The study that gave birth to the power tower; see especially Volume 4, "Economic Analysis."
2. "Barstow: Prototypical Power Project," A. C. Skinrood, in *Solar Age,* June 1978. A good report on the status of all aspects of the power tower program.
3. "Projection of Distributed-Collector Solar-Thermal Electric Power Plant Economics to 1990–2000," R. S. Caputo (Pasadena, CA: Jet Propulsion Laboratory, June 1977). A comprehensive analysis of the economics of fixed collectors, parabolic troughs, small dishes, and power towers for generating electricity.
4. "Semiannual Reviews of Solar Thermal Conversion Program Central Power Projects." SAND77–8011, Livermore, CA: Sandia Laboratories, March 1977). This and subsequent reports provide an ongoing account.
5. "Solar-Energy Conversion at High Solar Intensities," Charles E. Backus, in *J. Va. Sci. Technol.* Sept./Oct. 1975. An excellent short review of solar concentrating technologies that discusses the problems encountered at various intensities.
6. "Large Scale Central Receiver Solar Test Facilities," Giovanni Francia, in *Proceedings of the International Seminar on Large Scale Solar Energy Test Facilities,* Las Cruces, NM, November 1974. A good discussion of 100-kilowatt-sized power tower experiments, based largely on Francia's work from 1960 to 1968. (*Proceedings* available from H. L. Connell, New Mexico State University, Las Cruces, NM 88003.)

Chapter 3

1. "Photovoltaic Power Systems: A Tour Through the Alternativities," Henry Kelly, *Science,* 10 February 1978, p. 634. A thorough, up-to-date review article that deals with both technologies and systems; it has an extensive bibliography that provides entry to the technical literature.
2. "Photovoltaic Solar Energy Conversion," Martin Wolf, *Bulletin of the Atomic Scientists,* April 1976, p. 26. A useful introductory article now somewhat outdated by the rapid rate of technical development in photovoltaics.
3. "Novel Materials and Devices for Sunlight Concentrating Systems," H. J. Hovel, *IBM Journal of Research Development,* March 1978, p. 112. A good look at some of the future technologies for concentrating photovoltaic systems.
4. "Solar Cells Using Discharge-Produced Amorphous Silcon," D. E. Carlson, and C. R. Wronski, *Journal of Electronic Materials,* Vol. 6, 1977, p. 95. The basic details of one of the most promising thin film photovoltaic systems.
5. "Low-Cost, Low-Energy Processes for Producing Silicon," Lee P. Hunt, in *Semi-Conductor Silicon 1977,* H. R. Huff and E. Sirtl, Eds. (Princeton, NJ: The Electro-Chemical Association Soft Bound Symposium Series, 1977). A review of prospects for reducing the energy cost of silicon solar cells.
6. "Silicon Photovoltaic Cells in Thermophotovoltaic Conversion," R. M. Swanson, and R. N. Bracewell, EPRI ER-478 (Palo Alto, CA: The Electric Power Research Institute, February 1977). A description of an innovative new system for use with concentrating collectors.

Chapter 4

1. *Application of Solar Technology to Today's Energy Needs,* (Washington, DC: Office of Technology Assessment, 1978). This is the most comprehensive source of information on solar collector types, applications, and costs.
2. "Active-Type Solar Heating Systems for Houses: A Technology in Ferment," William A. Shurcliff, *The Bulletin of the Atomic Scientists,* February 1976, p. 30. An excellent introductory article.
3. *Direct Use of the Sun's Energy,* F. Daniels (New Haven, CT: Yale University Press, 1964). *Solar Energy Thermal Processes,* J. A. Duffie and W. A. Beckman (New York: John Wiley and Sons, 1974). These are the classic books.
4. "Simulation Analysis of Passive Solar Heated Buildings—Preliminary Results," J. D. Balcomb, J. C. Hedstrom, and R. D. McFarland, *Solar Energy,* Vol. 19, No. 3–E, 1977, p. 277. "Concrete Walls to Collect and Hold Heat," F. Trombe et al., *Solar Age,* August 1977, p. 13. These are among the few analytical studies of passive heating systems.
5. *The Solar Home Book,* Bruce Anderson and Michael Riordan (Harrisonville, NH: Cheshire Books, 1976). The best how-to-do-it book, this is a good introduction to practical aspects. More sources of practical information are listed in the "Solar Energy Bibliography" published by the National Center for Appropriate Technology, Butte, MT 59701.

6. *Solar Heated Buildings of North America*, William A. Shurcliff (Harrisville, NH: Brick House Publishing Co., 1978). Describing 120 solar-heated buildings (both active and passive), this is a complete and current guide. Description, cost, designer and builder, and performance figures are given for each house.

Chapter 5

1. "An Analysis of the Economic Potential of Solar Thermal Energy to Provide Industrial Process Heat" (Warrenton, VA: Intertechnology Corporation, February 1977). The best estimate of the solar industrial potential, the first volume (of three) is most valuable. Available from National Technical Information Service, Springfield, VA 22150.
2. "Report of the Solar Resources Group: A Supporting Paper for the Committee on Nuclear and Alternative Energy Systems (CONAES)" (1978). A thorough review that is particularly good on process heat. Available from the Printing and Publishing Office of the National Academy of Sciences, Washington, DC.
3. "Survey of the Applications of Solar Thermal Energy Systems to Industrial Process Heat" (Columbus, OH: Battelle Columbus Laboratories, January 1977). A widely quoted study that set the temperature limit for solar collectors unreasonably low (350°F) and derived a modest estimate of the contributions solar industrial energy could make.
4. "Concentrators: Collecting Total Energy," Robert Stromberg, *Solar Age,* February 1978, p. 8. A good short review of the Department of Energy work on total energy systems.
5. "Solar Thermal Power Systems Program" (Washington, DC: Department of Energy, Division of Solar Technology, January 1978). A summary of federal research contracts on total energy systems, irrigation systems, and power tower systems.
6. "Federal Agricultural and Industrial Process Heat Program Summary" (Washington, DC: Department of Energy, 1978). Contains the data that show how little is being done in the industrial field.
7. *Principles of Solar Engineering,* Frank Kreith and J. F. Kreider (New York: McGraw-Hill, 1978). Includes a chapter on process heat.
8. "Projection of Distributed-Collector Solar Thermal Electric Power Plant Economics to 1990–2000," Richard Caputo (Pasadena, CA: Jet Propulsion Laboratory, June 1977). One of the best analyses available of the cost of electric power from various types of on-site solar collectors.

Chapter 6

1. "Fuels from Biomass: Integration with Food and Materials Systems," E. S. Lipinsky, *Science,* 10 February 1978, p. 644. This and the next article are good reviews covering many aspects of biomass energy production and both include good bibliographies that are entrance points to the technical literature.
2. "Solar Biomass Energy: An Overview of U.S. Potential," C. C. Burwell, *Science,* 10 March 1978, p. 1041.

3. "Flower Power: Prospects for Photosynthetic Energy," Alan D. Poole and Robert H. Williams, *The Bulletin of the Atomic Scientists,* May 1976, p. 48. A good introduction.
4. "Silviculture Biomass Farms," R. E. Inman (The MITRE Corporation, McLean, VA: Report to the Energy Research and Development Administration, 1977, Vols. 1 to 6; also available from the National Technical Information Service, Springfield, VA 22150). This and the following reference are among the major recent reports dealing with biomass energy supplies.
5. "Crop, Forestry, and Manure Residuals Inventory—Continental United States," J. A. Alich, F. A. Schooley (Palo Alto, CA: Stanford Research Institute; Report to the U.S. Energy Research and Development Administration, 1976, Vols. 1 to 8; also available from the National Technical Information Service, Springfield, VA 22150).
6. "Synthesis-Gas Production from Organic Waste by Pyrolysis/Steam Reforming," Michael J. Antal, Jr. Paper presented at the Conference on Energy from Biomass and Waste, The Institute of Gas Technology, Washington, DC, August, 1978. This paper contains some of the most recent results on gasification of organic materials.
7. "Anaerobic Fermentation of Agriculture Residue: Potential for Improvement in Imitation," W. J. Jewell et al., Cornell University, Report to the U.S. Department of Energy, January 1978. This report contains some advanced results on new techniques for anaerobic fermentation.
8. *Biological Solar Energy Conversion,* Akira Mitsui et al., Eds. (New York: Academic Press, 1977). A useful compendium of international research efforts on the basic biology of photosynthesis and artificial photosynthetic systems.

Chapter 7

1. "Wind, Waves, and Tides," Marshal F. Merriam, *Annual Review of Energy* (to be published in 1978). An excellent short review of wind power and its status, with references that provide a good entry to the literature. A solidly based discussion of the prospects for wind use by Merriam, "Wind Energy Use in the United States to the Year 2000," is available from the Department of Energy, Energy Information Administration, Div. of Conservation and Renewable Resources, Washington, DC.
2. *Wind Machines,* F. R. Eldridge (1975, 77 pp.). A comprehensive review of the variety of wind machines, their costs, and applications—a very good introduction. Available from the Superintendent of Documents, Government Printing Office, Washington, DC 20402.
3. *The Generation of Electricity by Wind Power,* E. W. Golding (Philosophical Library, 1955; reprinted by Halstead–Wiley, 1976). A classic text.
4. *Power from the Wind,* P. C. Putnam (New York: Van Nostrand Reinhold, 1948). Good historical account of the 1200-kev machine Putnam built at Grandpa's Knob near Rutland, Vermont.
5. *Applied Aerodynamics of Windpower Machines,* R. E. Wilson and P. B. S. Lissaman, July 1974, PB–238–595. Complete and detailed discussion of aerodynamic theory. Available from National Technical Information Service, Springfield, VA 22150.
6. "Economics of Alternative Energy Sources," Martin Ryle, *Nature,* 12 May 1977, p. 55. Concludes wind is the most attractive alternative for Great Britain.
7. "Planning a Wind Powered Generating System," Enertech Corporation (1977). A good practical guide. Available from Enertech, P.O. Box 420, Norwich, VT 05055.

8. *Windpower Digest,* Michael Evans, 54468 CR31, Bristol, IN 46507. A quarterly that provides information and solicits grass-roots support.

9. "The Federal Wind Program, A Proposal for FY 1979 Budget," Rick Katzenberg and Ben Wolfe, The American Wind Energy Association, 1000 Connecticut Ave., N.W., Washington, DC 20036. This proposal concisely presents the case for an expanded wind program, and includes a selected bibliography that emphasizes government-sponsored work.

10. "Wind Energy Bibliography," published by Windworks, Box 329, Route 3, Mukwonago, WI 53149 ($3). An excellent bibliography to 1973.

11. "Energy from the Wind: Annotated Bibliography," B. L. Burke and R. N. Meroney, Colorado State University, Fort Collins, August 1975. First supplement, 1977.

Chapter 8

1. "A Proposed Conceptual Plan for Integration of Wind Turbine Generators with a Hydroelectric System," S. J. Hightower and A. W. Watts, March 1977. Available from the authors, Bureau of Reclamation, Department of the Interior, Denver, CO. The most thorough study of wind-hydro symbioses.

2. "Balancing Power Supply from Wind Energy Converter Systems," J. Molly, *Proceedings of the International Symposium on Wind Energy Systems,* Cambridge, U.K., September 1976.

3. "Wind Energy Statistics for Large Arrays of Wind Turbines (New England and Central U.S. Regions)," C. G. Justus, *Solar Energy,* Vol. 20, No. 5–B, 1978, p. 379. A study of the wind-averaging effect with many sites. See also J. Molly, *Wind Engineering,* Vol. 1, 1977, p. 57.

4. "Reliability of Wind Power from Dispersed Sites: A Preliminary Analysis," E. Kahn, LBL 6889 (Lawrence Berkeley Laboratory, April 1978). The wind fluctuation problem analyzed with a typical utility's loss-of-load models.

5. *Application of Solar Technology to Today's Energy Needs* (Washington, DC: Office of Technology Assessment, 1978). Chapter 11 is one of the best concise descriptions of the present status of storage research.

6. *Potential for Solar Heating in Canada,* K. G. T. Hollands and J. F. Orgill (Waterloo, Ontario: University of Waterloo, February 1977). Modeling solar energy which shows cost achieved with annual storage.

7. "Performance Report of the ACES Demonstration House," Eugene C. Hise, ORNL/CON–19 (Oak Ridge National Laboratory, March 1978). A description of an annual storage system, although one that uses an electric heat pump rather than solar power. Available from National Technical Information Service, Springfield, VA 22150.

Chapter 9

1. "Selected Issues of the Ocean Thermal Energy Conversion Program," A Report of the Marine Board, Assembly of Engineering, National Academy of Sciences, Washington, DC, 1977. A thorough review of the program and good guide to the literature.

2. "Renewable Ocean Energy Sources, Part 1: Ocean Thermal Energy Conversion," Office of Technology Assessment, May 1978. A review of the same issues as the Marine Board report, with similar conclusions. Available from the Superintendent of Documents, Government Printing Office, Washington, DC 20402.

3. "Perspectives on Implementing OTEC Power," David G. Jopling (1977). Candid discussion from a power company's point of view. Available from the author at Florida Power and Light Company, Miami, FL 33101.

4. Reports of Georges Claude: *Scientific American,* February 1930; *Mechanical Engineering,* December 1930, p. 1039; *Revue Scientifique* 28 March 1931, p. 161; *Chimie et Industrie,* January 1938, p. 3. These give a flavor of Claude's Cuban plant and his experiments with the ship *Tunisia* off Brazil, recording his successes and failures.

5. "Ocean Thermal Energy Conversion (OTEC) Power Plant Technical and Economic Feasibility Technical Report, Vol. 1." (Washington, DC: Lockheed Missile and Space Company, Inc., April 1975). The principal "mission analysis" for the program. Available from the National Technical Information Service, Springfield, VA 22150.

6. "Technical and Economic Feasibility of Ocean Thermal Energy Conversion," G. L. Dugger, E. J. Francis, and W. H. Avery (Baltimore, MD: Applied Physics Laboratory, The Johns Hopkins University, August 1976). Discusses the plant-ship concept.

INDEX

This book was set in Century Expanded,
Trade Gothic Bold and Video Medium
Oblique. It was printed and bound by
Waverly Press, Baltimore, Maryland.

Book design by Ellen Kahan
Production by Anne Holdsworth
Editorial coordination by Kathryn Wolff